Limits in Nature

Adopting a fresh approach, this book explores our understanding of nature's limits and proposes a new perspective to reframe the relationship between these limits and human freedom.

Where should we look to uncover the deeper roots of today's environmental disagreements?

A common answer might be: the conflict between nature's limits and human freedom. Yet, surprisingly, this critical topic has received little direct attention. To delve into this issue, the book turns to ontology; the philosophical study of being, and argues that our ontological understanding of nature profoundly shapes our ideas about freedom. By rethinking what limits in nature truly mean, we can gain new insights into what happens when we exceed those limits, into the real role of science and technology, and into what genuinely enhances freedom versus what diminishes it. Ultimately, the book contends that our current trajectory sacrifices the environment for a system that paradoxically offers less and less freedom. If this argument holds, it opens the door to reassessing the entrenched understanding of the conflict between human freedom and nature's limits to find a more lasting path forward.

This book is a must-read for environmental scholars seeking to deepen their understanding of the philosophical underpinnings of today's ecological challenges.

Svein Anders Noer Lie is an associate professor at The Arctic University of Norway—UiT in Tromsø. He has previously published the book *Limits in Nature: Rethinking Naturalness* (2016) and a number of papers in environmental philosophy, philosophy of technology and philosophy of medicine. He is the current leader of the environmental philosophy network at UiT.

Routledge Environmental Ethics
Series Editor: Benjamin Hale,
University of Colorado, Boulder, USA

The Routledge Environmental Ethics series aims to gather novel work on questions that fall at the intersection of the normative and the practical, with an eye toward conceptual issues that bear on environmental policy and environmental science. Recognizing the growing need for input from academic philosophers and political theorists in the broader environmental discourse, but also acknowledging that moral responsibilities for environmental alteration cannot be understood without rooting themselves in the practical and descriptive details, this series aims to unify contributions from within the environmental literature.

Books in this series can cover topics in a range of environmental contexts, including individual responsibility for climate change, conceptual matters affecting climate policy, the moral underpinnings of endangered species protection, complications facing wildlife management, the nature of extinction, the ethics of reintroduction and assisted migration, and reparative responsibilities to restore, among many others.

Virtue Ethics and the Environment
Dominika Dzwonkowska

Positive Duties to Wild Animals
Kyle Johannsen

Environmental Ethics and our Collective Character
David Clowney

Plant-Inclusive Care Ethics
Insights from Bonsai, Tomato Plants, and Giant Sequoia Trees
Kate Brelje

Limits in Nature
A Philosophical Exploration
Svein Anders Noer Lie

For more information about this series, please visit www.routledge.com/Routledge-Environmental-Ethics/book-series/REE

Limits in Nature

A Philosophical Exploration

Svein Anders Noer Lie

Routledge
Taylor & Francis Group
LONDON AND NEW YORK

earthscan
from Routledge

First published 2027
by Routledge
4 Park Square, Milton Park, Abingdon, Oxon OX14 4RN

and by Routledge
605 Third Avenue, New York, NY 10158

Routledge is an imprint of the Taylor & Francis Group, an informa business

British Library Cataloguing-in-Publication Data
A catalogue record for this book is available from the British Library

ISBN: 978-1-032-30647-6 (hbk)
ISBN: 978-1-032-30944-6 (pbk)
ISBN: 978-1-003-30737-2 (ebk)

DOI: 10.4324/9781003307372

Typeset in Times New Roman
by SPi Technologies India Pvt Ltd (Straive)

To my wife

Contents

Acknowledgments

I am grateful to several individuals for their invaluable support and contributions to this project. First and foremost, I extend my heartfelt thanks to Rani Lill Anjum for her meticulous and insightful reading of the manuscript. Her thoughtful comments and suggestions came at a critical moment in preparing this book for publication, reminding me of what true academic generosity and spirit look like. Secondly, I want to thank Yogi Hendelin for the enlightening and thought-provoking discussions, invaluable advice, and your remarkable scholarly breadth, all of which have significantly shaped this work. Pierce Stephens deserves special recognition for his assistance, particularly during the application process with Routledge. Thanks to the four anonymous reviewers at Routledge and what I learned from their insightful comments and criticism. I would also like to express my gratitude to my lifelong mentor, Finn R. Johannessen, and to Ove Aigner Haukenes for their readiness to engage in discussions and offer support and resistance whenever needed. I am also indebted to colleagues at the research seminar for their valuable feedback throughout the process. I should also mention Clare Heyward for her steadfast encouragement and support of this project, and Torje Mikalsen for our many enriching conversations. Thanks to Adele Ciriotti for co-organizing the workshop 'Limits in Nature—Can we Change Perspectives' with me, and to the other participants, Gaetano Chiurazzi in particular, from whom I learned a lot. Finally, I am sincerely grateful to our department (Tor Ivar Hanstad and Melina Duarte) for their support and the flexible accommodations provided at crucial moments.

Introduction

A great deal has been said and debated about freedom, yet comparatively little attention has been given to the subject of limits. In this book, I intend to explore how these two concepts and philosophical areas are interconnected. This idea has been developing for many years. At some point during my contemplation of this issue, it occurred to me that our understanding of freedom might not stem from a confined study of what freedom is in itself, but may have emerged from a particular understanding of limits in nature—specifically, from the larger cosmological and ontological reorientation occurring in the 17th and 18th centuries. This perspective convinced me that a philosophical exploration of limits in nature could potentially make a valuable contribution to the scholarly debate on environmental issues, as most environmental controversies ultimately boil down to the conflict between freedom and limits. What I present in this book is a more or less specific account and perspective on the relationship between limits in nature and freedom—a perspective that is supposed to challenge our current understanding of what limits in nature and freedom are. I am not presenting a comprehensive overview of an established research field, simply because this field does not exist. This has caused some concerns and difficulties along the way, as the intended audience for this book is not only environmental philosophers but environmental scholars more broadly. Since I cannot refer to an established research field, I need not only to introduce and connect various complex areas of philosophy but also make this accessible to a broader audience. The real difficulty in this case has been to find a way to fairly represent the positions I argue against. The questions I am asking are wide-reaching and I depend on quite general categorizations. What is our current understanding of freedom? What is the scientific consensus on how to describe nature? Is there a unified interpretation of 'ethics' and 'politics'? So, what allows for simplifications, and what requires deeper dissemination and exploration? I hope I have been able to strike a balance between these dimensions and that the text is both accessible to a wider audience, interesting to environmental philosophers and fresh enough to make a real contribution to the scholarly debate on environmental issues.

We often take for granted the necessity of self-limitation to avert environmental collapse or degradation. Thus, we see that those who prioritize (conventional) freedom above all else engage in the environmental debate through

DOI: 10.4324/9781003307372-1

a spectrum of denial: environmental problems are rejected, doubted, minimized, and, if acknowledged, believed to be solvable through technological progress. Thus, to limit ourselves for the sake of the environment is viewed as a strategy to subordinate people under socialist rule, or something similar. On the other side, we have (progressive) environmentalists who are confused by a conundrum. All the successful progressive struggles of past centuries have revolved around themes of freedom and liberty. Peasants, slaves, women, workers, nations and minority groups have had their time of liberation against oppression and the 'conservatives' who sought to uphold illegitimate power structures. But what about environmentalists? Are they freedom fighters? Do they sit on the right side of history? They believe the time has come to liberate nature. But wait, isn't much of the freedom and welfare we enjoy today a result of our progressive liberation from nature? Isn't the liberation of nature forcing us to limit our own freedoms? Wouldn't large portions of us be enslaved and deprived of our birthrights if it weren't for our mechanical tools' taming of nature? If this is the way we see it, Robert Green Ingersoll's Indianapolis speech (1876) serves as a vivid illustration:

> I am a Republican because that party believes in free labor. It believes that free labor will give us wealth. It believes in free thought, because it believes that free thought will give us truth. You do not know what a grand party you belong to. I never want any holier or grander title of nobility than that I belong to the Republican party, and have fought for the liberty of man. The Republican party, I say, believes in free labor. The Republican party also believes in slavery. What kind of slavery? In enslaving the forces of nature.
>
> We believe that free labor, that free thought, have enslaved the forces of nature, and made them work for man. We make the old attraction of gravitation work for us; we make the lightning do our errands we make steam hammer and fashion what we need. The forces of nature are the slaves of the Republican party. They have no backs to be whipped, they have no hearts to be torn—no hearts to be broken; they cannot be separated from their wives; they cannot be dragged from the bosom's husbands; they work night and day and never tire. You cannot whip them, you cannot starve them, and a Democrat even can be trusted with one of them (…)
>
> Slavery never invented but one machine, and that was a threshing machine in the shape of a whip. Free labor has invented all the machines. We want to come down to the philosophy of things.
>
> (Ingersoll 1876)

The quote from Ingersoll's speech may seem outdated. As we have achieved a broad level of comfort and welfare in many Western societies, we now seem to enjoy the luxury of viewing 'nature' not as a counterpart to our existence and interests, but rather as a victim and a space 'where humans can become themselves.' From this perspective, the power of Ingersoll's words has faded and may even seem "Mordorish." However, this is not the complete story, as we are simultaneously exploiting and destroying what is not strictly human on this earth at an unprecedented rate. The deep currents remain with us (Marx, 2000). Is this because there will always be a categorical conflict between human freedom and the limits imposed by nature, or is what we are witnessing now a hyperbolic expression of the internal inconsistencies inherent in, for example, Ingersoll's thoughts? I argue that the latter is the case. Our current *modus operandi* is not rational; it does not lead to greater freedom and is not motivated by what was originally intended. It is a system running on a recipe gone wild. It is a system whose relationship between freedom and limits seriously needs to be reconsidered from its philosophical roots.

I am not the first to claim this or to address this issue. So, what is distinct about the outlook and scope in this case? In one word? Ontology. Ontology is the branch of philosophy that tries to deal with the question, "What is real?" In contemporary philosophy, many consider this question contentious. I am not one of those. One reason is that I believe this skepticism is based on the idea that we cannot say anything about this question before we know how to access it, which is, I argue, itself based on an ontological claim (Section 2.10 and Lie, 2025). Another reason is that we cannot avoid adopting some kind of idea of what we think is real when we act and think. For instance, there are ontological assumptions underlying our ideas about freedom and limits. Unaware of such assumptions, we are less able to understand and address ethical and political issues in ways that can make a difference, if and when that is needed.

The claim in this book is that our frontier culture, which threatens to undermine the conditions for life, originates from a particular ontology I call a part-limiting cosmology. Humans have flourished under this cosmology, but at some point, problems began to emerge behind our backs. My primary concern in this book is the environmental issues on one side and the steady advancement of a whole-limiting system on the other. In the latter case, I am not referring to the rise of totalitarian states, socialism, artificial general intelligence (AGI) or the concentration of tech power. Instead, I am discussing a "physically" integrated system, a whole that constrains the freedom which the part-limiting cosmology was originally conceived as a counterpart to. We find ourselves in a situation where we are destroying the natural world around us without gaining any additional (conventional) freedom in return. This is a conundrum that even the most libertarian environmental skeptic must come to terms with.

There is a particular understanding of limits *in nature* that allows for the argument that 'entrenched freedom' and environmental problems share the same root. This understanding, which I term a relational-limiting cosmology, may also give rise to a viable alternative to the frontier culture. This book aims to explore this alternative understanding, which not only provides a framework for rethinking limits and freedom but also offers different perspectives on technology, the nature/human divide, ethics and human goal-directedness, the role of politics and the nature of innovation.

References

Ingersoll, Robert G. (1876). Indianapolis speech. In *The Works of Robert G. Ingersoll*, vol. 9. New York: Dresden Publishing Company, 1900.

Lie, S. A. N. (2025). Making sense of 'description from within': An ontological challenge to (eco-) phenomenology. In M. T. Heneise et al. (Eds.), *The Routledge handbook of ecophenomenology*. Routledge.

Marx, L. (2000). *The Machine in the garden: technology and the pastoral ideal in America*. Oxford University Press. (Originally published in 1964).

1 The question of limits in nature

1.1 Introduction

Limits are everywhere. There are limits to how high one can jump, how fast a Boeing 777 passenger plane can fly, how long water can remain in a liquid state when temperatures drop, how long a liver cell in a cod can survive, how much an aquifer can be depleted, and how high temperatures a leopard can endure, among others. Even if we don't address these limits or think about them, they are still there and some might even say they constitute our being.

Imagine the following scenario. From today, you will rid yourself of the trappings of civilization and go live in the woods. Let's assume that the woods where you will live are situated in a temperate, continental climate zone. You might have to drive your car to get there, but at this point, you must leave it behind. From now on, you will survive without any aids or supportive tools. For this, you should also remove your shoes and clothes. As you venture further, bare and unguarded, you will start to encounter limits: the limits of the human body, the limits of comfort and other constraints imposed by nature. At this raw juncture, you will experience the limits of your human existence. Without shoes to protect your feet from the rough ground, you can no longer walk as freely as you once did. Sleeping at night might become increasingly difficult, and feeding yourself might become critically challenging, depending on the season. All your activities will be heavily reliant on weather conditions. How long will it take before you have to change your circumstances? Is it likely that you will be able to change your situation without retreating all the way back to the civilization you left behind? What would you need to do, and at which point would you give up and say enough is enough?

In order to survive out in the wild, without the aid of modern-day means, a number of adjustments are needed. You would need to become more independent of the weather conditions, for instance, and you would need some form of housing and clothing. You would need more energy to sustain yourself as a body. You would need more knowledge about your environment, and perhaps some additional technological devices to help you gather and cook your food. In a temperate and continental climate zone, seasons vary considerably, which means that you would need techniques to store food through seasons

DOI: 10.4324/9781003307372-2

with little access to food. If this is not enough, you might need to trade goods for other goods, transgressing the limits of your place. To do this, you would need to engage with other people and be able to reach agreements and trust. Local limits of nature may then become regional limits of nature, regional limits may become national limits of nature, and national limits may become global limits of nature.

At which point do you end up in our current situation, where most of us think of limits in nature as global limits? (cf., e.g., The Planetary Boundaries Framework). There is a wide gap between the tangible limits you are bound to experience in the example and the global limits we are mostly worried about today. We live in a world where limits in our personal life seem to be just a matter of money, personal choice and technoscientific breakthroughs, while there is little direct experience of the global limits which seem to be at the center of the limits-debate today. Do they exist at all? Many of us might even think that 'limits in nature' is not a very trustworthy notion, as expressed in sayings like 'limits are a product of our mind,' 'endings are just new beginnings,' or 'the sky is the limit.' Our economy runs on the assumption that we can grow and expand forever.

Those who have warned about limits to certain non-renewable resources have been proven wrong again and again, and are often disrespectfully called Malthusians (Kallis, 2019, 11 and 21). Our experience and expectations seem to be that science and technology will manage to find ways to overcome limits. Indeed, limits that were more or less considered 'real and absolute' are routinely overcome and celebrated as important breakthroughs. Gazing into the "unlimited" universe, it is also difficult to even imagine that the limits we are talking about are real.

This situation gives those who argue that we must learn to impose limitations on ourselves a hard time. We see that politicians advocating for restrictions on air travel, the expansion of business opportunities (that harms the environment), the consumption of more goods, etc., are perceived as proponents of diminished freedom. They are seen as leftists or more generally, 'people who use the environmental cause to propagate their own gloomy outlook on humanity'. The idea that there is a conflict between limits in nature and our freedom is also accepted by many scholars who focus on the themes of societal restrictions and personal self-restraint and other virtues. Scholars who think the other way around worry that '(L)imits straightjacket political deliberation and action, undemocratically turning public life into ecological management by experts. In short, limits are inconsistent with liberty, whether conceived as liberty from external interference, or liberty as collective democratic self-governance and self-determination.' (Cannavò, 2020, 82).

Despite these overarching disagreements, the question of limits does not seem to go away. There is a new call for planetary boundaries, postgrowth, degrowth, ecological ceilings and carbon footprints, to mention only some (Kallis, 2019, 3). Last time this type of debate was vibrant was in the 1970s. The question is: have we moved forward from the last time limits were debated?

Do we have the right starting point to address these issues, or do we first need to reassess our understanding of what 'freedom' and 'limits are'? Is it true that 'being limited' and 'being unfree' are synonyms: the less limited, the more free?

In what follows, I will try to make the case that the objective of our current civilization is not freedom, not even in its conventional sense, but rather the transgression of limits as a principle. The mantra seems to be that what can be done, should be done: 'can' entails 'ought', irrespective of any genuine concern for freedom. I would call this an irrational expansion and as this expansion happens in the name of limits vs. freedom it may also be easier to accept that there is a real opportunity to reassess our understanding of what limits in nature and freedom are, and how they relate.

1.2 Frontier culture

Where there are expansion and growth, limits are called into question. Those concerned about our expansionist behaviors toward allohuman beings—such as ecosystems, habitats, species, and the planet itself—are often labeled as environmentalists. Typically, environmentalists focus their efforts on raising awareness and promoting action: If we acknowledge these limits, action will follow. For instance, *if* climate change is a real threat, and *if* we should not exceed a concentration of 430 parts per million (ppm) of CO_2, *then* action will follow. This is then the expectation: Science identifies the limits, and politics responds by managing our activities accordingly. While this may be a simplistic representation of how we address the question of limits today, it is difficult to find evidence of a deeper understanding in the public discourse. For instance, most research categorized as 'environmental problems' is conducted within the natural sciences, mapping the consequences of our expansion in ever more detail—while downplaying the need to understand the social drivers of that expansion. Is it truly the case that the limits we face result from mere rational choice, and can therefore be resolved by a rational acknowledgment of the need to limit ourselves? While it may be true that 'limits are back,' we do not seem to care much about finding the right context to understand, act, and behave in ways that do not reinforce what we are trying to avoid.

First of all, despite longstanding criticism of the idea of progress, we do not seem to properly recognize our embeddedness in a frontier culture. Previous cultures, such as the Ancient Greek culture, found it important to come to terms with the limits of our human lives (*sophrosyne*). In contrast, the sentiment of our modern Western culture seems to be one of challenging all kinds of limits. Transcending limits for its own sake is considered a meaningful activity. For example, in science, we would not view it as a problem if scientists suddenly declared they would stop exploring beyond what we already know; rather, we would see it as meaningless. Our economy, as it functions today, would not only come to a halt if we decided to stop growing; it would collapse. What would happen to Apple if there were no new versions of their phones? The virtues of being inventive, creative, and original—once associated with the avant-garde in art—now permeate society. Mass avant-gardism dominates the Western world. It is flooding the world by its constant creative destruction, disruptive technologies, and transgressive operations that question everything and deny ourselves nothing: shouting into the indifferent universe, "Anything Is Possible."

A country stops at its borders. The frontier culture resembles an *empire* in that it only halts where resistance is strong enough to limit further expansion. 'The Empire' has afforded everything human in the Western realm. Thus, it appears contradictory to argue that *there are* limits in nature, since those limits are merely the points where counterforces are sufficiently strong to restrict further growth—precisely what is constantly transgressed down the line. Limits seem to be nothing in themselves, just arbitrary points on a timeline. Does this present a situation where our best hope lies in believing that more scientific knowledge about these limits will lead to relevant political actions?

A longstanding criticism of this naïve stance posits that our expansion is a product of a certain modern ideal that this very stance embodies. In a recent book, David McPherson (2022) refers to this trait in our culture as the Promethean ideal, a term that has circulated for decades. Prometheus, the Titan who stole fire from the Olympian gods and gave it to humanity, suffered eternal punishment from Zeus. The myth serves as a warning to those who challenge the gods, but in our case, it illustrates an ideal where we are enabled to exceed our limitations and predicament through a choosing-controlling mindset, as McPherson describes it.

The question is: Does this mindset accurately depict our current reality post modernity? I would argue that the 'choosing-controlling mindset' ideal is fading.[1] The expansion of our frontier culture today appears more blind, wild, and irrational. Science is in the business of developing technologies that are intentionally uncontrollable[2]; few would argue that our economic system serves 'the good life'[3] and ethics/politics seem ill-equipped to rationally manage how we undermine the conditions for further biological life on Earth. (Frodeman, 2019, e.g., xf). Our recent experience with AI exemplifies this impression, as even the *developers* express doubts about whether this technology is truly controllable (Becker, 2025).

Furthermore, there seems to be no common goal defining progress. 'Breaking limits,' 'expanding into the unknown,' 'being flexible,' 'being creative and open-minded,' have become ends in themselves (Bauman, 2000). The driving forces of society—economy, science, technology, and art—appear to be in a fluid, transitional state, increasingly reliant on politics and administration for coordination and maintenance. A dependency that complicates politics' ability to remain visionary.

A fair objection would be to say that this is not new. Marx and Engels expressed this already in their Communist Manifesto: 'All that is solid melts into air' (Marx & Engels, 2012, 38). There is however a difference in that the upheaval, and 'constant revolutionizing of the means of production' was a by-product of 'the bourgeoisie's pursuit of profit and dominance', not an end in itself. Now, these are values expressed as ideology. Constant disruption, transcendence, flexibility and dynamism are virtues, not a by-product of a goal that *can be* criticized. Importantly, the point here is not to complain about this situation as such ('stop the world, I want to get off'). The point is that one has to make a choice. Either you argue that this process of expansion and transgression of limits leads to more freedom (and accept that this means that the 'dynamic development' must be tamed by a choosing-controlling mindset), or you bite the bullet and say that 'this is how it is and must be: What we are talking about is our destiny and not a pursuit of freedom and happiness, and thus beyond criticism and evaluation.' You cannot have it both ways.

In the latter case, the best justification seems to be to say that 'our cultural state is an expression of human biological instincts—our control and rational choice have always masked the deeper biological forces anyway and this is what we have now unleashed in our modern "de-biologized" way of life.'[4]

In this book, I offer a different analysis. Besides the fact that the term 'biological instincts' relies on the dualistic Enlightenment idea of control and rational choice (being its antithesis), it is also misleading. (See Section 2.5). Rather, I believe we can argue for an analysis that connects our expansive frontier culture with a *certain* idea of freedom. I see our frontier culture as a completion of this idea.

In my view, what we need is therefore to resurrect a proper concept of limits that does not take *our* current concept of freedom as a given. To be more precise: When we discuss our 'unlimited expansion,' I believe we also tacitly accept a certain perspective on what limits really are. Hence, naively arguing that these (type of) limits are real in an effort to counter optimists or social constructivists will not only mislead us but also reinforce our 'unlimited expansion', which is what is taking place when tech-millionaires like Jeff Bezos and Elon Musk invest in the idea that our only solution would be to expand into space to occupy our solar system and beyond (Becker, 2025, 23). For those who don't believe in eternal economic growth in the first place and thus await some kind of breakdown, it can be useful to explore a different idea of limits and freedom. This may also help in the process of making the right choices if the time comes.

1.3 What I *can* do and what I *should* do

Our frontier culture seems to have reached a point where what we can do is equated with what we should do. From a Promethean 'choosing-controlling mindset,' this perspective is problematic. We should be able to control our actions and choose what we should do independently of what we *can* do. The issue arises because this view only makes sense if we can maintain a fairly strict distinction between 'is' and 'ought,' between science and politics, freedom and ethics, humans and the natural world, and so on.

Several reasons contribute to the problematic nature of this perspective. I have written extensively about the ontological issues associated with this distinction in Lie, 2016 (see especially Chapters 1.3–1.4), and I will provide a broad sketch of some of these problems relevant to the current case.

1.3.1 Positive and negative freedom

Most people in Western countries live in a liberal democracy where we differentiate between positive and negative freedom. Negative freedom refers to what we can and cannot do. When someone is deterred from doing something they want to do, they are not free. The only restriction accepted by liberalism is when one's freedom harms another's freedom (the harm principle). Positive freedom, on the other hand, adds substance to 'I want.' What 'I want' is not merely a fact about the individual's desires, as one may want something that does not actually enhance their freedom. Our understanding of freedom requires qualification beyond 'I want to do x'. For example, some might argue that a person may freely spend their life on social media, yet this could lead to a situation where they are ultimately less free. If we all shared the same understanding of what it means to be human—a common (philosophical) anthropology—we could agree on what we should truly want.

However, from a liberalist perspective (negative freedom), such an agreement is not possible and would additionally foster a totalitarian society (see Isaiah Berlin, 1969). The meaning of 'leading a good life' *is* subjective. Therefore, there should not be a common understanding limiting what we should do. The only consensus we can reach pertains to what we can and cannot do. (See also Sections 2.5 and 6.3).

While this is a well-known problem, it is often overlooked that we should not operate *as if* we still have the power to restrict 'can' by 'should.' If we in fact act solely based on what we can do (though limited by the harm principle), the issue lies not in a lack of ethics, public debate or international agreements, but in the frontier culture itself as this steadily provides a rising plethora of what we can do.

1.3.2 Science is no longer pure science

Science has also changed. The notion of science as a neutral provider of true knowledge is no longer tenable, especially as scientific institutions have become more systematically integrated into our competitive economies, transitioning from an industrial to a knowledge economy (Powell & Snellman, 2004). This shift was already noted in the 1990s (Funtowicz & Ravetz, 1993; Gibbons et al., 1994; Ziman, 1996). Science is no longer 'pure science'; it is significantly involved in shaping society in specific directions. Science occupies a unique position: its core pursuits are viewed as free intellectual inquiries, yet science also functions as an instrument for continuously revealing what we can do. In other words, science not only answers what we can do but also partly "suggests" what we should do (cf. Sections 4.6 and 4.9). Consequently, what we should do is *primarily* constrained not by what we can do but by economic considerations (capital/investment) (see Section 6.6): Expanding our capabilities is beneficial as it stimulates the economy, which is advantageous for humanity. If we fail to act on what we can do, we risk falling behind in the economic race, jeopardizing our relative competitive status in the global economy (cf. the race to develop artificial general intelligence [AGI]).

1.3.3 There are ontological strings attached

The idea of what we can do is heavily influenced by a particular ontological view of nature. The prevailing passivist perspective reduces the question of 'can do' to a matter of causality.[5] While we may agree for various reasons to exclude humans from being categorized as part of (passivist) nature and accept that there *are* additional ethical limits on how we treat other people, it becomes challenging to justify any ethical considerations for other parts of nature. This leads us to a conceptual vacuum. Limits in nature that extend beyond what we perceive as causally possible *are, in this context, seen as limits that must be ethical to count as limits*; however, since we cannot ascribe any ethical value to other parts of nature (as Aristotle's teleology might allow), the only limits we should accept are those defined by science. This implies that there are no limits in nature that inform us about what we should do, only limits on what we can do. These limits are essentially what we consider to be the 'laws of nature.' Hence, if this assertion holds true, there is little need for an elaborated discourse on limits in nature.

1.3.4 We have physical limits and we have socially constructed limits?

An alternative argument posits that there are indeed limits in nature—both biological and nonbiological—when we consider the consequences (utility) for ourselves. For example, if we wish to hunt grouse in an area over an extended period, there is a definitive limit to how much we can reduce the current population. Similarly, if we aspire to live a specific lifestyle, there is a limit to how much additional carbon we *should* allow into the atmosphere. Therefore, we might regard the nine planetary boundaries identified by the Stockholm Resilience Centre not as limits in nature, but as constraints on *our* activities.[6] This perspective aligns with our current understanding of limits: In addition to 'real limits in nature' (e.g., the speed of light), there are certain properties of nature that restrict humans whenever *we seek to* achieve a specific goal, which we *should* not exceed for that reason. Limits seem to be relative to the goals we have. However, if we accept these 'conditional limits', we also imply a different understanding of what limits really are. We accept what I call a relational understanding of limits.

Clarifying the implications of this understanding is the primary subject of this book. If all limits are indeed relational, including those that coincide with the supposed laws of nature, we may also acknowledge that they exist in nature. If limits are relational, they are neither external nor internal; they exist in nature as part of a historically evolved relational order, of which *we* are also a part. The argument presented in this book posits that the limits I will refer to as 'limits in nature' (or: 'historically evolved relational limits') *are not* contingent. They are relational, not relative. This means they cannot be bent, manipulated or surpassed without reprisal.

Thus, the main contribution of this book is grounded in the idea of a certain *order* in nature that may support a specific type of 'nonethical ought.' While this might seem like a minor shift, professional philosophers will recognize its significance. Discussing a specific order in nature that transcends the order implied by the purported laws of nature raises various controversial issues. Are we reverting to the axiological cosmologies of thinkers like Aristotle and Aquinas? Are we confusing the orders created by our mind with a purported order of the world? Does this necessitate abandoning the is/ought distinction? Do we require a form of politics that is not solely justified by reference to the voice of the people?

The starting point here offers numerous pathways that may lead to entrenched positions. The task of this book is to find a way to navigate out of the entrenched corner we are currently facing, while providing an account that may promise future developments. The challenge is to demonstrate that there are limits *in* nature that are amenable and not fixed, yet still real, and which, to a certain extent, offer advice on what we should do.

The main counterargument to the view I explore here is the perspective that (real) limits are only those that are fixed. These typically include limits such as the speed of light (in a vacuum: 299,792,458 m/s) and the absolute freezing point (−273.15°C). Those that are not fixed are deemed unreal limits. Discussing them as real invites accusations of constructivist activity. They are often lumped together with limits that serve various (power) interests, such as 'women should know their limits,' 'slaves should know their place,' 'nonfigurative painting is not art,' and 'don't walk on the lawn.' I consider this dichotomy between real (fixed) limits and constructed limits to be a faithful servant of our expansionist frontier culture, where we freely construct the world according *to our* more or less considerate subjective 'oughts.'

1.4 The optimist and the pessimist

The problem described in the previous chapter can have an easier solution if we assume that our progressing manipulations of nature will lead to a better human life. If nature represents limits to our freedom, then transgressing those limits will, *prima facie*, represent something good. If this is the case, we don't need to consider whether we should do what we can do. This is the idea of progress, which dates back to the Enlightenment (Idhe, 1993, 61. See also Section 2.4). The concept of progress serves as a proxy for the Aristotelian teleology that faded away simultaneously: when you expand into your telos (i.e., 'progress'), you are guaranteed to also follow the good. Should is entailed by can.

Many would recognize that one of the more explicit roots of the idea of progress goes back to Francis Bacon (1561–1626). In *The New Atlantis* 2002, Bacon famously visualized how the 'inquisition of nature' could occur in what he called Solomon's House, where the secrets of nature could be captured, controlled and transformed. The aim of this house was not only to interrogate nature but also to convince the public that such investigations were not sinful, thereby wresting this "scientific activity" from the ill-reputed hands of the secretive alchemists (see Merchant, 2015, Chapter 4). The idea was that if the public recognized that the ultimate aim of this interrogation was to provide humans with the means to control and manipulate nature for human welfare, the associated superstition would fade.

Another root stems from the notion that humans are not at home in nature and will only become *themselves* through the 'humanization of nature' and not, for instance, through religious belief. This idea was further developed by thinkers like Georg Wilhelm Hegel, Karl Marx, and Ernst Kapp (see Mitcham, 1994, Chapter 1).

A more longstanding root comes from a certain Christian tradition that sees humans as liberators of nature. Nature will become true through humanity, the pinnacle of creation. Wilderness or "raw nature" was viewed as the remnants left behind by the Flood, where the descendants of Noah were tasked with restoring creation to its original state. Bringing civilization also meant cleaning up the untidy nature of the uncivilized (Nash, 1967). Even as Christianity faded and faced challenges from the Enlightenment, this idea complemented the broader concept of progress, with progress and Christian civilization often going hand in hand during European colonialism.

Most importantly, the idea of progress was obviously also fostered by its own success. The new ways to reorganize and rearrange nature provided jobs, prosperity, safety and the prospect of equality.

The pessimist, on the other hand, is skeptical of the entire enterprise of progress and questions whether the humanization of nature through science and technology makes us more human. The pessimist would argue the opposite: it leads, in the first instance, to a new type of alienation (Karl Marx).[7] It results in the instrumentalization of rationality, turning the prospects of enlightenment into myth (Theodor Adorno & Max Horkheimer). It entrenches a perspective that selectively reveals beings of nature – and even humans – as resources (Martin Heidegger and Jacques Ellul). It transforms our human interests into consumer interests driven by amusement and instant gratification (Herbert Marcuse, Daniel Bell, Neil Postman). It misguides us regarding what is truly important in our individual lives (Albert Borgmann). Many others could be added to this list of critics.

A more general criticism, however, is that the optimist lacks a clear idea of what the humanization should serve. There is no clear understanding of what humans are. Most optimists find this question unsettling. Isn't it clear what we all want? We want to be happy. We want to free ourselves from toil and sorrow. We want a healthier, longer life. And anyway, if you want something else, in a liberal democracy, you are free to pursue most of your desires if you put in the corresponding effort.

The pessimist would respond that we cannot determine whether there is progress without a unified idea of what humans truly are. Without a certain understanding of this, optimism will thrive under any conditions. The optimist would still be an optimist in a Blade Runner world. The ongoing humanization of nature will be blind, and assuming that progress is good is not wrong, but meaningless. Without an idea of what humans are, there would be no limit to what technologies can do to us and to nature. The pessimism here is not about whether progress will lead to a better situation, but whether progress is a meaningful term.

Whether this pessimism is well-founded or if there is a direct connection between our humanization of nature and the failure to create a "true human society" is, of course, a matter of debate with a lot of moving parts. What seems clear, however, is that there is little indication that this process can or will be stopped or morph into something else. The pessimist's critique of the humanities does not seem to make a difference either.[8] Given that we seem stuck in a motion toward an infinite transgression of limits in nature (initially), it does not really take much to be a pessimist. To raise the question of limits suffices. One might therefore assume that the only thing able to limit the expansive humanization of nature is nature itself. But will 'reality strike back'?

The reader has probably anticipated the following: If the optimist cannot believe in limits and does not subscribe to an idea of human nature, there is little hope for arguments. This is why many environmentalists have trusted in some form of 'reality effect,' where the environment 'punishes' the idea of progress (cf. Lovelock, *The Revenge of Gaia* 2007). They believe nature will 'speak back' in a clear voice. Yet this notion seems to favor a notion of fixed limits again. That they are out there – stopping us. So, if that does not happen, it seems the optimist will continue to thrive. This is why it is important to understand what limits truly are. Without an alternative conception of limits, we will not know how limits arise, how and when they stop us, if that is indeed what is happening.

It should not be surprising, therefore, that a philosophical study of limits must include a better understanding of how freedom and limits are intertwined. The only way to demonstrate that limits are real is to show that further progress leads to less freedom of the kind we seek when we call for more progress.

Before I delve into the basic conditions underlying the relationship between limits and freedom, we need to discuss more concretely how limits have entered the public debate today. We need to clarify the concept of limits currently being used and normalized.

1.5 Limits-discourse and sustainability-discourse: Limits are back

Limits are back. (See, e.g., Lysaker 2023, 7.) What does that mean? Following the surge of the Great Acceleration, critical voices began to emerge during the 1960s and 1970s. Some of these voices came from economic scholars, with the most notable publication, The Limits to Growth Report (LGR report) by Meadows & Meadows and Randers (Meadows et al., 1972), quickly becoming a focal point of debate. Harsh criticism followed, such as:

> It's imposing apparatus of computer technology and systems jargon conceals a kind of intellectual Rube Goldberg device—one which takes

> arbitrary assumptions, shakes them up and comes out with arbitrary conclusions that have the ring of science. "Limits" pretends to a degree of certainty so exaggerated as to obscure the few modest (and unoriginal) insights that it genuinely contains. Less than pseudoscience and little more than polemical fiction, "The Limits to Growth" is best summarized not as a rediscovery of the laws of nature but as a rediscovery of the oldest maxim of computer science: Garbage In, Garbage Out.
>
> (Passell et al., 1972)

The authors of the LGR report were not satisfied with the ensuing debate. In fact, they were shocked by the attacks launched against the report. They believed they had presented a set of rational arguments that could be discussed appropriately. However, very few actually addressed the metrics, the models, and the scientific merits of the report.[9] The received impression of the report was difficult to change, since the criticism was rather general. Although the claims in the report were 'conditional' (if, then), they were accused of presenting limits as absolute and fixed. Critics argued that science and technology were not considered. Thus, what may seem like a limit at one point could later prove to be an unfounded concern. "(E)xtrapolation of the trends of the 1880s would show today's cities buried under horse manure." (Passell et al., 1972).

A strong argument against the flawed interpretation of limits in the report—as something 'out there'—was made by Julian Simon in his book The Ultimate Resource (1981), where he argues that resources are not found but invented: Nature, in terms of matter, does not possess economic value; it gains value only when put to use by humans. For example, oil seeping from the ground was merely greasy and unpleasant until humans learned to convert it into energy. Similarly, sand was just dusty and hard to walk on until someone transformed it into glass. Nature serves as a raw substrate for transformation, acquiring value through our use. Just as Aristotle believed that matter is meaningless without form, Simon suggests that humans can give form to matter in ways that differ from nature's own processes. Knowledge is the key factor; it is our ultimate limit, as it restricts the ways we can reshape matter and "unlock" resources.

A similar framework was posed by Eric Drexler, who according to Ed Regis was motivated by the Limits to Growth Report when he coined the notion nanotechnology (Regis, 1995, 87).

> Beef was protein, and proteins were combinations of amino acids. Amino acids, in turn, were nothing but molecules of ordinary elements – carbon, hydrogen, oxygen and nitrogen, bound together in certain stable configurations. Cows made beef by placing the required molecules into the necessary configurations. Drexler's meat machine would do the same thing. (….) The meat machine would be a mechanical cow, a factory at the level of atoms.
>
> (Regis, 1995, 7)

'There's Plenty of Room at the Bottom,' Richard Feynman argued in a 1959 lecture at the California Institute of Technology (2011). Therefore, to assume that resources x, y and z are limited and will constrain future economic and population growth is a misunderstanding of what resources are.

Additionally, we do not need to plan how to navigate and surpass the limitations on growth at any given time; the economy will largely handle this. Established economies will always move away from scarcity as prices rise, and new innovations will overcome what we currently view as obstacles to further development (Sandbach, 1978, 505). Modern economies focused on technological development will ensure that technological *efficiency* compensates for the rise in economic activity (decoupling).

When comparing these responses to contemporary perspectives, we will see that ecomodernists and trans-humanists argue along similar lines (Cannavò, 2020, 83–86). Ecomodernists believe in decoupling and welcome the idea that 'We are Gods now, and we better be good at it' (Stewart Brand). Nature has a close to endless plasticity and we can remake whatever nature offers to our own benefit, although within certain limits that science prescribes. Transhumanists would agree to this but argue that we should not only recreate our environmental conditions but also go a step further, to recreate human beings. If the environment becomes uninhabitable, we can adjust to these conditions as well. Ultimately, there is nothing stopping us from adapting our human biology to a Blade Runner world, if we want or have to.

We also recognize these positions from leading figures grown out of Silicon Valley, such as Peter Thiel, Elon Musk, Marc Andreessen, Ben Horowitz and others (see, e.g., The Techno-Optimist Manifesto).

While there might be valid and general arguments against some of the premises in the LTG report, many would also recognize that the challenge of economic growth is that technological efficiency must keep pace with a geometric (exponential) growth curve. This is articulated in the I=PAT equation developed by Erlich and Holdren around the same time as the LTG report. The equation states that environmental impact equals population multiplied by affluence (growth) multiplied by technological efficiency. The global population was approximately 7 billion in 2009. The average income per capita (affluence) was around $8,000.[10] One thousand dollars corresponds to about 0.5 tons of CO_2 emissions in 2009, leading to total emissions of 28 billion tons annually. Assuming 450 ppm is the limit for emissions before triggering uncontrollable climate change (e.g., Hothouse Earth), we arrive at the following estimates.

The human population is projected to reach about 9 billion by 2050. The growth in population would reduce our individual footprint to 0.1 tons per $1,000, which seems achievable if we consider technological efficiency. However, factoring in a 3% annual growth rate drastically alters the situation. Technology would then need to improve efficiency 11-fold, necessitating a 98% increase in efficiency compared to 2010, by 2050. By 2070, we would need to maintain the same efficiency rate, and again by 2080.[11] These figures highlight the point made by the LGR report, which also probably explains why this was unsettling news in the 1970s for the rising middle class, who had enjoyed a 'well-deserved increase' in income with an even brighter future ahead. It might be wiser to heed the critics rather than dampen the celebration.

In this situation, the Brundtland report became useful to effectively stall the debate on limits to growth. Instead of addressing the problems related to the exponential growth curve of our economy, the report focused on human needs. From this perspective, it seems reasonable to require economic growth to enhance global equality, but within the framework of what was famously termed 'sustainability,' where a development is sustainable 'when it meets the needs of the present without compromising the ability of future generations to meet their own needs' (WCED, 1987, 43). The shift from "external limits in nature" ('object') to our internal "nature" ('subject') also reflects a broader sentiment that had been developing for some time, culminating in the heydays of postmodernism and social constructionism in the 1990s. Scientific realism and the notion of nature as something definable were viewed as overly naïve. This change in *stimmung* also came to the rescue of the broadening of conspicuous consumption to the general population taking place in the 1980s.[12] As the critiques of the LGR report gained traction (Sandbach, 1978), one could argue that the limits in nature pointed out by the LGR report largely vanished. Where did nature go?

One might expect it to reemerge as a counterpart to 'human needs.' However, if the limit to the gratification of my needs is set by the needs of future generations, we must agree on a universal definition of needs that corresponds to specific *requirements* of our natural environment. To reach that agreement would necessitate defining what it means to be human, which to a certain extent contradicts the principles of liberal democracy (see Sections 1.3 and 6.3). It also conflicts with the neoclassical economic definition *of needs, which equates needs with preferences* (O'Neill, 2010; Toman et al., 1994, 402). Additionally, it is philosophically challenging to distinguish between real and artificial needs or simply needs and wants (O'Neill, 2010 and e.g., Cutts, 2019).

But if we cannot really pin down limits in nature in terms of needs, it seems that the whole idea (of limits) has vanished. In some sense, this might be true. In another, it is also true that this question found its way into neoclassical economic theory: Supporting needs now and in the future means supporting a certain level of aggregated income (measured in GDP). It makes no difference whether that income comes from natural or human capital.[13] We can destroy a natural system as long as some of the capital gained is reinvested to grow or maintain a certain level of aggregated income (Hartwick's rule). If not reinvested, it is not sustainable. If the satisfaction of future generations' needs (preferences) is the limit, then reinvestment becomes the only criterion required for a sustainable society. Limits in nature only emerge when the destruction of the natural world prevents an increase in aggregated income through reinvestment. But when does that happen?

Neoclassical economic theory appears to have all the necessary tools to overcome this problem. It can translate natural capital into (growing or sustained) financial (human) capital, manipulate our needs (demands in a market), commodify solutions for limits, move investment resources quickly, and make the organization of production, commerce, transport, and societal institutions more efficient. Nowadays, it also seems capable of changing and channeling our values and political priorities through digital operations. A fractured nature is not a problem but a challenge. Pushing beyond current limits is seen as a business opportunity (viz., today's talk about 'the green transition'). This is how Luigi Pellizzoni (2021, 84) describes it:

> As storylines like sustainable development, ecological modernization, and green economy indicate, in the post-Fordist regime of accumulation, environmental issues become economic opportunities. Instrumental to this account is a new scientific thinking, developing alongside in chemistry, cybernetics, and life sciences, which, contrary to earlier assumptions, sees normalcy in disorder, open-endedness, patchiness, fragmentation, competition, and unpredictable dynamics, deemed to bring renewal and change against the 'heat death' of equilibrium (Cooper 2008, Walker and Cooper 2011). In this framework, material limits are simultaneously acknowledged and pushed forward. Technological and organizational innovation is assumed to translate them into contingent barriers to production. (….). The case for limits to growth is reversed into a case for the growth of limits.

The more problems there are, the more business ideas emerge. We are not settling into a state of harmony or a final endpoint of human flourishing. Instead, we are in a continuous process of becoming through destruction and reconstruction—existentialism on a societal level. While sustainability looks like a model of harmony and balance, it really morphed into a supportive agent accentuating the idea of unlimited expansion, according to Pellizzoni.

The LGR Report offers a broader, more scientific critique of the notion of progress that previously came from 'pessimist philosophers' and other theorists with limited audiences. This has helped optimists refine their arguments and clarify their vision of progress. Progress is not a goal-directed process moving toward paradise and control; rather, it is an ongoing challenge and conflict where all parts are dynamically in motion. However, for this narrative to hold, the parts *must* truly move beyond the models.

The assertion is that they do not. Environmental problems have not diminished; rather, the opposite is true. To a large extent, the use of matter, energy, and the production of pollution/waste follow the economic growth curve, and the projections in the report were surprisingly accurate.[14] The current resource use (2016) amounts to 1.69 times the Earth's carrying capacity (The Global Footprint Network). The Stockholm Resilience Centre does not shy away from discussing planetary boundaries, claiming that 7 out of 9 have been crossed (2025). The question remains whether highlighting these seemingly unmovable parts will finally persuade optimists to change their approach.

The LGR report was correct in pointing out limits to growth but failed to emphasize the right understanding of what limits truly are. As will be duly argued in later parts of this book, limits are relational. They exist in nature but are not external to the agency of individuals or collective bodies. Their *particularity* depends on what we want to do and who we want to be. In this sense, Julian Simon and the report on sustainability were right to assume that these limits depend on our desires. No needs, no limits. Many needs, many limits.

However, the problem is that many limits represent numerous business opportunities. The expansion of limits goes hand in hand with market growth, leading to more growth and more limits, and so on. Drought caused by climate change creates a market for plastic lawns, longer fire hoses, and bottled drinking water. Our interaction with nature as a collective body is through economic activity. The limits of nature are economic limits (Section 6.6), which means that limits in nature do not appear as a single number in the equation. Rather, they appear as a number that quickly becomes a plethora of relational numbers. For example, we can emit a certain amount of greenhouse gases by 2050, but *if we do x*, we can emit a different amount.[15] In an article published in *The Conversation*, James Dyke, Robert Watson, and Wolfgang Knorr provide a brief introduction to how these integrated assessment models work in practice. One of the examples is when the US negotiated to include forests as carbon sinks in the equation.

> Around the time they were first developed, efforts were being made to secure US action on climate by allowing carbon sinks from the country's forests to count. The US argued that effective forest management would enable it to store significant amounts of carbon in trees and soil, which should be deducted from its obligations to limit fossil fuel use. In the end, the US largely got its way.
>
> (Dyke et al., 2021)

Once limits in nature are converted into economic models, we can, by adding and subtracting different agents (e.g., forests as carbon sinks) into the equation, play with different scenarios and reach very different conclusions about limits in nature (e.g., on climate emissions). Few openly claim that there are no limits in nature, but many find it reasonable to follow the conclusions reached through the infinite flexibility of our economic models. The assertion that we cannot have infinite growth in a finite world is not reflected in economic models because limits are relational, which is also what is assumed in economical models, although for the economist, relational means contingent (Section 6.6).

The economy is about the future and if we do x, y and z, we will learn that what occurs as a problem now is not really a problem later on. Money is debt to the future, and we can be made to believe in many things about the future. Investment relies on varying levels of belief, shaped by past experiences and a steady stream of positive news from the science and technology frontier. It takes considerable effort to dismiss the idea that a problem is larger than the business opportunity it presents. While this might not surprise those familiar with the field, it is crucial to clarify, as it highlights that those attempting to argue for limits in nature outside of 'economic laws' will find themselves outside the current discourse.

As long as human-induced climate change poses a problem that can be translated into healthy returns for investors—through reinvestment in renewables, carbon-sink systems, security and climate-adjusting initiatives—it seems irrational to advocate for drastic emission cuts that require deeper systemic changes in the economy, culture, and society. Demonstrating that we have crossed seven out of nine planetary boundaries fails to influence those who do not value ecosystem preservation (Kingsnorth, 2013). A functional system will suffice (Lie, 2016, Chapter 6.4). The consequences of cutting down the Amazon rainforest may be severe, yet it could also generate aggregated income for future generations, especially if the problems created by this destruction lead to products and services that foster new growth. As long as the repercussions can be absorbed into the commons without significantly impacting *isolated* investment opportunities arising from these issues, economists may see no reason to incorporate (fixed) ecological limits into their models. Although challenging, it is not impossible for the political economy to distribute this aggregated income in a manner deemed just and satisfactory to the populace. The only thing that does not make sense economically is if we do something that is not economically viable. Our interface with nature as a collective knows only one world, and that is the world of contingently related parts. Economic instruments serve to combine, tear apart, and restructure these parts, including humans.[16]

Those who trust the flexibility that the economy provides in our current situation should probably also hold onto the new type of optimism that celebrates creative destruction, 'wild and disruptive technologies,' and 'setbacks as a pathway to future prosperity.' (See Nordhaus & Shellenberger, 2007, quoted and discussed in Cannavò, 2020, 84.) The only worry for such a scenario seems to be when the cycle of investment and reinvestment is disrupted. This is the case with downward spiraling feedback loops (potentially caused, e.g., by negative feedback mechanisms in the climate system) where the complexity makes it impossible to track and map 'causes and effects' in the (global) system, thereby preventing the ability to foresee the outcome of an investment.[17]

While this worry may be warranted, there is a much more interesting limit that I care about in this book. If we view limits in nature as being relational, there is a different type of problem that occurs. I have argued that a "relational view" of limits is evident in how (neoclassical) economics works its way around problems. However, on an ontological level, this approach reduces relationality to mere relativism and contingency. This means that those who follow the optimist scenario described above are unable to see how our growing manipulations of nature create a whole-limiting system that requires top-down governance of its parts, including humans. Thus, the parallel growth of such systems is blamed on left-leaning politics. If limits were seen as truly relational, we would understand that limits of nature are never really overcome. They are just expressed differently. And these expressions are not a result of ideological disagreement. Moving related parts around as if they were contingently stacked Lego bricks cannot be done for free. It requires realigning work and continuous maintenance that leads to ever more complex human-made systems. On a relational account of limits, freedom does not come down to a simple struggle to overcome impediments.

In the chapters to come, I will work out an account of what 'relational limits' means and what I see as the true implications of such a view.

Notes

1 … and that this has been going on for a while. An influential, deeper and earlier critique of this mindset was put forward by Horkheimer and Adorno in *Dialektik der Aufklärung* (Horkheimer & Adorno, 1972), but this was already indirectly challenged (or maybe complemented) by Daniel Bell (1976) and others.

2 Bongard and Levin (2021).

3 Cf. The Easterlin Paradox.

4 The idea that humans are bound to expand or transcend can be established by pointing to our biology or a particular anthropology like transhumanism. See, e.g., Bostrom, 2005, 1.

5 For a far more detailed description of passivism, see Lie (2016, chapter 2), and Sections 2.4 and 2.8.

6 See also Raworth, 2017, where limits of nature (ceiling: 9 listed) define the 'safe space for humanity.' The human activity that Raworth sees as necessary is defined at the center of the doughnut. I.e., we *could* choose to change our idea of what it means to lead a good life and adjust the limits accordingly.

7 Marx was undoubtedly an optimist but leaned his optimism on a more radical idea of what it takes for progress to succeed. The productive forces may align with the relations of production, unleashing the potential of the forces in a certain time period, but that does not mean that humans are *freed from alienation*, which is the very rationale behind the humanization of nature. The idea of progress does not reach its endpoint before alienation is abolished, and a state of (Hegelian) recognition rules the relationships between people.

8 Cf. Robert Mitcham, 1994. Mitcham lists many of the critiques that I mention in passing above as a 'humanist critique'. The reason why I claim that this criticism does not seem to make a difference is not easy to substantiate beyond doubt. What I can say is that the impression stems from my readings and teachings of philosophy of technology. In my view, the critiques listed above in passing are well founded, thorough and to a certain extent true. Instrumental rationality *is* prevailing. The technological development *is* in fact becoming more and more 'autonomous.' Despite this, it does not seem to make an impact, not even within the area of philosophy of technology. Leading figures within this field today are proponents of the empirical turn and want to reject the pessimism entailed in the immanent critique by studying technology as social constructions, which to me seems to be an escape from further discussion rather than an honest attempt to find ways to improve the understanding of technology. See, e.g., Bijker, 2013; Feenberg, 2016; Nordmann, 2016; Winner, 1993.

9 https://donellameadows.org/archives/the-history-of-the-limits-to-growth

10 These numbers are taken from *Prosperity Without Growth* (Jackson, 2009).

11 There have been given many similar calculations through the years since the LTG report. An updated and convincing example which also addresses more recent issues raised by ecomodernists and transhumanists would be Thomas Murphy's paper *Limits to Economic Growth* from Murphy (2022).

12 A change that was precisely anticipated in the classical work, *The Cultural Contradictions of Capitalism* (1976), by Daniel Bell.
13 In neoclassical economic theory, natural capital can be *substituted* for human capital, which is viewed as weak sustainability. Strong sustainability is defined by the nonsubstitutability of certain natural goods such as eco-services, the integrity of ecosystems, species, etc. (Nöel & O'Connor, 1998).
14 https://en.wikipedia.org/wiki/The_Limits_to_Growth (see heading: 'positive reviews' and Nebel et al., 2023).
15 In the case of limits to climate change emissions, this is called 'Integrated Assessment Models (IAMs),' which eminently describes how limits in nature are translated into economic limits in economic models (Yang et al., 2020).
16 'Including humans.' What I mean by adding this is to again stress that limits in nature *also* depend on a more or less fixed understanding of what it means to be human and what a good life looks like. Humans are an extremely flexible species and can be led by all sorts of ideas of a good life. The economy is now increasingly getting a hold on instruments to manipulate such ideas (e.g., AI and social media), which was previously a reserved area for religion, humanities/social science and public debate (ideally expressed through rational discourse).
17 A different example of a messed-up situation like this is when we cannot really trace the causes of a person's death. A person who starts with a health problem solved by a medicine that creates new issues, which in turn creates further problems while exacerbating the original issues, is "solved" by new medicines that generate additional problems, and so on until death.

References

Bacon, F. New Atlantis. (2002). In Brian Vickers (Ed.), *The major works* (pp. 457–489). Oxford: Oxford University Press. Originally published in 1627.

Bauman, Z. (2000). *Liquid modernity*. Cambridge: Polity Press.

Becker A. (2025). *More everything forever: AI overlords, space empires, and Silicon Valley's Crusade to control the fate of humanity*. New York: Basic Books.

Bell D. (1976). *The cultural contradictions of capitalism*. New York: Basic Books.

Berlin, Isaiah (1969). Two concepts of liberty. In *Four essays on liberty* (pp. 118–172). Oxford: Oxford University Press.

Bongard, J., & Levin, M. (2021). Living things are not (20th century) machines: Updating mechanism metaphors in light of the modern science of machine behavior. *Frontiers in Ecology and Evolution*, 9, Article 650726. https://doi.org/10.3389/fevo.2021.650726

Bostrom, N. (2005). A history of transhumanist thought. *Journal of Evolution and Technology*, *14* (1). http://jetpress.org/volume14/bostrom.html

Brundtland Commission (WCED). (1987). *Our common future*. Oxford University Press.

Cannavò, P. F. (2020). Limits and liberty in the Anthropocene. In C. J. Orr, K. Kish, & B. Jennings (Eds.), *Liberty and the ecological crisis*. Routledge.

Cutts, J. (2019). Herbert Marcuse and "false needs". *Social Theory and Practice*, *45*(3), 353–370. https://www.jstor.org/stable/45237362

Dyke, J., Watson, R., & Knorr, W. (2021). Climate scientists: Concept of net zero is a dangerous trap. *The Conversation*, April 22. https://theconversation.com

Feenberg, A. (2016). The concept of function in critical theory of technology. In F. Fransen, P. Vermaas, & P. Kroes (Eds.), *Philosophy of technology after the empirical turn* (pp. 245–260). Springer. https://doi.org/10.1007/978-3-319-33717-3_16

Feynman, R. P. (2011). There's plenty of room at the bottom. *Reson*, *16*, 890–905. https://doi.org/10.1007/s12045-011-0109-x

Frodeman, R. (2019). *Transhumanism, nature and the ends of science*. Routledge.
Funtowicz, S. O., & Ravetz, J. R. (1993). Science for the post-normal age. *Futures*, *25*(7), 739–755. https://doi.org/10.1016/0016-3287(93)90022-L
Gibbons, M., Limoges, C., Nowotny, H., Schwartzman, S., Scott, P., & Trow, M. (1994). *The new production of knowledge: The dynamics of science and research in contemporary societies*. Sage Publications.
Horkheimer, M., & Adorno, T. W. (1972). *The dialectic of enlightenment*. Trans. J. Cumming. Herder and Herder. (Original work published 1947).
Idhe, D. (1993). *Philosophy of technology: An introduction*. Pragon House Publishers.
Jackson, T. (2009). *Prosperity without growth: Economics for a finite planet*. Routledge.
Kallis, G. (2019). *Limits: Why Malthus was wrong and why environmentalists should care*. Stanford University Press.
Kingsnorth, P. (2013). Dark ecology. *Orion Magazine*, January/February. https://orionmagazine.org
Lie, S. A. N. (2016). *The philosophy of nature: Rethinking naturalness*. Routledge.
Lovelock, J. (2007). *The revenge of Gaia*. Penguin Books.
Lysaker, O. (2023). *Ecological democracy: Caring for the Earth in the Anthropocene*. London and New York: Routledge. https://doi.org/10.4324/9781003305842
Marx, K., & Engels, F. (2012). *The communist manifesto: A modern edition*. Verso.
McPherson, D. (2022). *The virtues of limits*. Oxford University Press. https://doi.org/10.1093/oso/9780192848536.001.0001
Meadows, D. H., Meadows, D. L., & Randers, J. (1972). *The limits to growth*. Universe Books.
Merchant, C. (2015). *Autonomous nature: Problems of prediction and control from ancient times to the scientific revolution*. Routledge.
Mitcham, C. (1994). *Thinking through technology: The path between engineering and philosophy*. University of Chicago Press.
Murphy, T. W. (2022). Limits to economic growth. *Nature Physics*, *18*(8), 844–847.
Nash, R. (1967). *Wilderness and the American mind*. Yale University Press.
Nebel, A., Kling, A., Willamowski, R., & Schell, T. (2023). Recalibration of limits to growth: An update of the World3 model. *Journal of Industrial Ecology*. https://doi.org/10.1111/jiec.13442
Nöel, J. F., & O'Connor, M. (1998). Strong sustainability and critical natural capital. In I. S. Faucheux & M. O'Connor (Eds.), *Valuation for sustainable development* (pp. 75–98). Edward Elgar.
Nordhaus, T., & Shellenberger, M. (2007). *Breakthrough: Why we can't leave saving the planet to environmentalists*. Houghton Mifflin Harcourt.
Nordmann, A. (2016). Changing perspectives: The technological turn in the philosophies of science and technology. In F. Fransen, P. Vermaas, & P. Kroes (Eds.), *Philosophy of technology after the empirical turn*. Springer.
O'Neill, J. (2010). The overshadowing of needs. In F. Rauschmayer, I. Omann, & J. Frühmann (Eds.), *Sustainable development: Capabilities, needs, and well-being*. Routledge.
Passell, P. et al. (1972). Review of the limits to growth. *The New York Times Book Review*, April 2.
Pellizzoni, L. (2021). Nature, limits and form-of-life. *Environmental Politics*, *30*(1–2), 81–99. https://doi.org/10.1080/09644016.2020.1868864
Powell, W. W., & Snellman, K. (2004). The knowledge economy. *Annual Review of Sociology*, *30*, 199–220. https://doi.org/10.1146/annurev.soc.29.010202.100037
Raworth, K. (2017). *Doughnut economics: Seven ways to think like a 21st-century economist*. Random House.
Regis, E. (1995). *Nano: Remaking the world atom by atom*. Bantam Press.
Sandbach, F. H. (1978). The rise and fall of the limits to growth. *Social Studies of Science*, *8*(4), 495–520.

Simon, J. (1981). *The ultimate resource*. Princeton, NJ: Princeton University Press.
Toman, M. A., Pezzey, J. C. V., & Krautkraemer, J. A. (1994). Neoclassical economic growth theory and "sustainability". In D. Bromley (Ed.), *Handbook of environmental economics* (pp. 139–165). Blackwell. https://www.jstor.org/stable/284820
Bijker, W. (2013). The social construction of technology. In J. K. Friis, S. Pedersen, & V. Hendricks (Eds.), *A companion to the philosophy of technology*. Wiley-Blackwell.
Winner, L. (1993). Upon opening the black box and finding it empty: Social constructivism and the philosophy of technology. *Science, Technology & Human Values*, *18*(3), 362–378. https://doi.org/10.1177/016224399301800306
Yang, Y., et al. (2020). Integrated assessment models for climate change. *Oxford Bibliographies*. https://doi.org/10.1093/obo/9780199363445-0043
Ziman, J. (1996). Is science losing its objectivity? *Nature*, *382*(6594), 751–754. https://doi.org/10.1038/382751a0

2 Philosophizing limits

Introduction

Before we can reach for an alternative view on limits and freedom, we need to explore and understand the philosophical background of how these concepts are generally understood today. I have proposed that our frontier culture can be attributed to 'our human nature,' which drives us to expand indefinitely *or* alternatively to a specific and historically contingent understanding of what freedom and limits really are. I will support the latter assumption, which is also what I will aim to justify throughout this chapter by exploring what I call the part-limiting cosmology. I contend that this justification must be philosophical. Our concepts of limits and freedom are inherently philosophical, and at the moment, it is hard to see how we could argue against 'the inevitability of our expansionist nature' without turning to philosophy.

Some may protest and say that it is more relevant to examine the economy, suggesting that capitalism explains what we need to know. However, an analysis of capitalism does not advance our understanding of what limits are; rather, it relies on our existing notions of limits and freedom. Capitalism *assumes* boundless expansion, but this assumption is not founded on a theory of nature and limits. Instead, it is our general theories about limits and freedom that make capitalism possible and viable. Additionally, there has been no shortage of theoretical and ideological contributions that point to capitalism as a significant reference point for those seeking to understand our unbounded activities on Earth. Perhaps it is time for a deeper philosophical exploration that explicitly examines our ideas about limits themselves?

2.1 Parts and wholes

Limits play a crucial role in one of philosophy's most fundamental questions: the relationship between parts and wholes.[1]

If there are no parts in the world, can there be limits? If there are parts, how do we distinguish between them, and how do they relate to the whole? Are parts constituted by the whole, or is the whole merely a product of the parts? Do parts simply manifest the whole, as Plotinus, Spinoza and some early Greek philosophers suggested? If I possessed superpowers that allowed me to do anything I wished, would I be a part or the whole itself? If I am the unlimited

DOI: 10.4324/9781003307372-3

whole, do I have parts and if so, am I constrained by them? Would my will be subjected to any limitations? If I could do whatever I wanted, would there ever be a conflict between my wants? Would there be something that made one action conflict with another? If I am a part, how would I be limited by the other parts and by the whole, if there is such a thing as The Whole? If I am a part, in terms of 'taking part,' would I be able to discern where I begin and end? If I am a part, is it possible for me, from an external perspective to see its boundaries? If parts are not organized by the whole, can they still be ordered (and limited), or would they simply exist in arbitrary groupings? Are the parts self-contained essences, and if I, as a part, were to exceed my limits, would I become something entirely different? These are some of the philosophical questions that arise when we consider the part/whole issue in relation to limits.

I will argue that the discussion of limits has been significantly shaped by our understanding of the relationship between parts and wholes. Historically, it is pretty clear that limits were originally perceived within what I call a whole-limiting cosmology.[2] In this view, *parts are mainly viewed as limited by the whole.* This view shifts during a period of perhaps 2–300 years, from the 17th to the 20th century, and one may say that we are probably still in the midst of this transformation. This "new" perspective, which I refer to as part-limiting cosmology, describes parts that exist independently of any whole. When the manifest belief in a particular whole disappeared, the 'being part of' also vanished. (Which is, I would suggest, the deeper meaning behind the phrase 'God is dead.') In a part-limiting *cosmology, parts are primarily understood as limited by other* parts. Our contemporary understanding of limits is largely influenced by this view, where we see ourselves as parts among other parts, rather than as components of a unified whole. This perspective has significant implications for our understanding of freedom and limits.

As an alternative to both whole-limiting and part-limiting cosmologies, I propose a 'relation-limiting cosmology.' In this framework, *parts are mainly viewed as being limited by their relationships to other parts* (and sometimes by the whole that the relational dynamics constitute). In this chapter, I will explore whole-limiting and part-limiting perspectives, while the relation-limiting view will be discussed in Chapter 3.

Imagine two trees growing side by side: the whole-limiting tree and the part-limiting tree. Though they are distinct opposites, their roots beneath the surface agree; they depend on their differences to maintain their status as the only two trees available to climb.

2.2 There is a whole of which x is a part

In an attempt to challenge the traditional way of thinking with the scope of Greek mythology, early Greek philosophers asked themselves, 'Why is there order (cosmos) rather than chaos?' The bare question implies that chaos is the more likely state of existence. What stands in need of an explanation is therefore cosmos, not chaos. Chaos would be the state of unformed matter where everything changes continuously without any order or pattern. Already from its very inception, the idea of cosmos clearly favors a way of thinking that situates parts within a defined whole, of which x is a part (Copleston, 2003, 76). The hierarchical primacy of cosmos places everything *within* this order, where nothing can escape it. Hence, the inscription over the entrance to the oracle of Delphi: 'Know thyself.' Failing to know your place and exceeding your limits (hubris) would bring about nemesis. Breaking with the order of the universe is bad; staying within limits is good. This applies not only to humans but to every part. While this may reflect a trait of Greek culture in general, the tenet became more explicitly articulated during the prime of Greek philosophy.

The early Ionian pre-Socratic philosophers famously began their inquiry by asserting that there must be some kind of 'ur-stuff' or first principle (arche) underlying the changes we observe. (Aristotle, The Metaphysics 983b). This principle orders and explains the semi-chaotic state of the world. Thales argued that water was the best candidate; Heraclitus, fire; Anaximenes, air; and Empedocles, all four elements.

Anaximander argued that all beings spring from the boundless or limitless (*to apeiron*), which has no beginning or end (Aristotle, *Physics* 203 b6–b15). The boundless "awaits" limits to be imposed upon it to become 'something' (see Aristotle, *Physics* 1996, p. 247, comment to note 203 a4).[3] According to 'the (incalculable) measure of time,' beings are manifested and given a specific being through limitations and then are drawn back into the boundless (circle of creation and decay). According to Anaximander then, *limits* are the creative force of the *particular* and manifest world as we know it. Beings are not *something* that *eventually* becomes limited. Aristotle interprets Anaximander as suggesting that if anything becomes limitless, it will destroy the rest; thus, beings (in this case, the elements) must differ from the limitless but also 'come into being from it' (Aristotle, *Physics* III.5 204 b 22). Limits are at the very center of being and creation.

The fact that beings are what they are in virtue of being limited does not provide much insight into the specific order of things, except to indicate that if something could be or do *everything*, it would cease to be.

When considering the pre-Socratics and Greek philosophy in general, we must keep in mind that our modern concept of causation as an efficient cause is not particularly fitting. Today, we tend to view the (natural) world as an arrangement of universal cause-and-effect processes, where a cause produces the same effect across time and space, is prior to the effect in time, and can be clearly separated from the effect. For example, light (photons) causes carbon dioxide and water to produce glucose and oxygen in the presence of chlorophyll.

The Greek word for cause (aitia) stems from being 'guilty of something.' Hence, void, water, fire, logos and apeiron can be responsible for something without causing it in the modern sense. The different archai's do not explain the order of things in terms of *how* something happens, but *why* (notably without involving gods, which would revert back to Greek mythology). The arche is a general principle that governs everything, and this principle is responsible for bringing beings into existence. In this sense, the idea of arche is not very different from mythological thinking (which the pre-Socratics sought to move away from) if we consider the gods as descriptions of behavior. When Poseidon is described, one gains a broader understanding of the nature and dynamics of the ocean's realm. Unlike mythological thinking, this approach aims to uncover an abstract principle that governs all things. However, it is somewhat challenging to conclude that this means the principle is the whole of which x is a part if the part is categorically different from the whole, as in Anaximander's case.

One may conclude that the Greeks generally did not think that the parts of the world are determined and limited by their "antecedent causes," but rather by the principles that guide their behavior. In Plato and Aristotle, these principles are differentiated once again. In Plato, the principles are *ideas*; in Aristotle, *form*, which he calls 'a principle of rest and change.' How the different parts of the world are limited to behaving in certain ways is explained in both cases by a more general principle: the principle of the good.[4] Everything there is has an inner drive toward becoming what it is. This is why we refer to the ontology of Plato and Aristotle as 'teleological ontology.' The drive (*entelechy*) is a striving toward perfection, and striving toward perfection is a striving toward the good.

2.3 Part/whole and teleology

A teleological cause exemplifies a situation in which the whole is a real entity, with X as a part rather than the other way around (i.e., where the whole is merely an aggregate of its parts). For something to possess a teleological cause, it must be a whole that directs its parts toward a specific goal. This goal is an end, which describes a limit to 'how far you can go.' You cannot go beyond this end. Being a part of this whole, you are a part of this end. In Aristotle's view, for instance, the whole (substance) directs its parts (matter) toward an end determined by its form (Aristotle, *Physics* 193, a8 ff.). According to Aristotle, I am composed of matter—specifically organs (organon means tools) in the case of living beings—but it is not the organs (parts/matter) that 'is guilty' of my movement; rather, it is I, as a whole organism, that possesses a telos. The nature of this whole is defined by my form. My identity as human, as opposed to being a blackbird, is not determined by my parts (matter) but by my human form. The whole exerts a form of "downward causation" (to use an anachronistic term) that compels the organs to function in a manner conducive to the realization of the organism's form. It is the organism that has the potential to experience fear (e.g., of an approaching soldier) and mobilizes its organs to 'address' the challenges perceived as relevant (e.g., engaging muscles to pick up a gun).

What limitations do teleological causes impose on us according to Aristotle and Plato? According to both, 'nothing more' than privation. I may act in ways that cross my "principle of rest and change" or fail to "find my place." There is no "whole-limiting force" that *directly* restricts my actions. A disease or a person plagued by vices represents a deviation or imbalance. My limitations intensify if I disregard guidance from my form/idea, leading to a state of being 'more' limited (e.g., sick, overcome by vices or even monstrous). To transcend limits signifies a failure to fulfill one's potential. In contemporary terms, going beyond limits could be likened to what the Greeks termed hubris. Transgressing these limits inevitably leads to suffering (*nemesis*).

In discussing biological beings, Aristotle argues that a (real) substance, composed of form and matter, can become matter (part) within a higher-level substance. For instance, a heart consists of both form and matter but functions as matter when viewed as (qua) part of the organism as a whole. The heart serves a functional role for the organism "*for* the sake of which" it exists. In other words, we see that in the organism, the parts are subordinated to the whole (form) of the organism.[5]

But what about global or cosmic natural teleology? Is there an overarching whole with a telos of which we and other 'mid-sized' substances are functional parts? This question presents a puzzle for Aristotle (Aristotle, *Physics* 198 b16). He agreed with the materialists of his time that it would be odd for rain to fall solely for the purpose of corn growth, as it would also lead to the conclusion that rain falls for the purpose of making corn rotten (from excessive moisture). But he does not think the falling of rain to be entirely contingent or random. The general principle is that everything occurs for the sake of something; everything happens 'by nature,' meaning it is due to a teleological cause (Aristotle, *Physics* 198b16–199a8). But what does this imply in terms of the cosmic whole? While it is necessary for rain to fall when cooled and for the water cycle to function as it does, is this due to an inherent telos of the water element or to plant growth? Is there an "extrinsic" vertical order of which we, along with plants, are a part?[6] One might argue that Aristotle's recognition of this as a real puzzle prevents him from formulating a comprehensive theory of a cosmic telos, unlike Thomas Aquinas. (For a discussion on this issue, see Rossi, 2010, 6.)

For sure, Aristotle does not begin with the idea of a cosmological telos, inferring from this that the parts must derive their cause from such a cosmological telos. In fact, that would imply that we would see that his systematic study of "the nature of things" (substances) first and foremost would emphasize the 'in-order-to' of mid-sized substances like us. Rather, he perceives mid-sized beings as having an immanent form where the telos is inherent, not *primarily* for the sake of something else (Johnson, 2005, 4 and 186; Rossi, 2010, 7). The actions of a plant, for example, in realizing its potential are not guided "top-down" for the sake of a higher end; instead, they are driven by its own *intrinsic* end. The form exists in the *substance independently of everything else*. While starting with mid-sized parts might allow him to infer something about the whole, he does not approach this in a systematic or detailed manner (see Rossi, 2010, 8 (2)). Kant, two millennia later, would discuss a similar topic in a similarly cautious manner in his *Critique of Judgment*, § 67.

When contemplating a united whole, we typically envision a large substance, a physical unit existing in space and time. This is not how Plato conceptualizes it. Plato thinks that we participate in *nous*, a higher reason or mind.

In *Timaeus*, Plato recounts the story of the Demiurge (the Craftsman God), who shapes the ordered world (cosmos) as we perceive it through his reason, based on the invisible eternal forms that resonate through the infinite 'realm of the ideal' (Mason, 2013, 203).[7]

The crafting process resembles that of a human artisan: arranging everything into a mutually supportive whole that is ultimately good (Johansen, 2020, 32 ff.). The concern for the whole prevails: "Even if one might theoretically consider a part of a visible work which has no intelligence more beautiful than some part of a visible whole with intelligence, the whole with intelligence will be more beautiful than the other whole" (Johansen, 2020, 34). A part is primarily "motivated" by the pursuit of the good, which cannot be defined independently of the whole. Thus, the whole (the principle of nous) is 'guilty of' the (mid-sized) part and its telos. However, the part is not "instructed" by the telos; rather, it is a component of the whole in the sense of participating in nous (e.g., Phaedo, 100d; Plato, 1997). This principle organizes everything and serves as their cause.

> (I) was delighted with this cause, and it seemed to me good, in a way, that mind should be the cause of all. I thought that if this were so, the directing mind would direct everything and arrange each thing in the way that was best. If then one wished to know the cause of each thing, why it comes to be or perish or exist one had to find what was the best way for it to be, or to be acted upon, or act.
>
> (Phaedo, 97c–d)

The Demiurge created the world as we know it based on eternal forms and did so by virtue of nous, which permeates the world.

> The Demiurge serves as the condition for Plato's revival of the *nous* doctrine, acting as the transcendental guarantor of the goodness of the world soul and cosmic self-regulation. Nevertheless, it is not the Demiurge that governs the world, but a *nous* that is immanent within the world itself. Plato also needs this immanent nous; his need for a moral universe is not fulfilled if morality is solely supplied by divine intervention from outside.
>
> (Mason, 2013, 210)

As noted, the Greek term cosmos signifies an ordered whole. Plato believes this order is governed by nous, which creates a good order wherein the parts are directed, *also* limited by other parts, toward the right end (Mason, 2013, 208).

While there are differences between Plato and Aristotle, one may wonder if these differences truly matter in our case. In one sense, they do not, as the parts in both philosophers' natural philosophies behave as if they are part of a whole. They are not first and foremost unorganized parts that are 'up against' other parts but are rather their own measure sticks. In Plato's case, their measure sticks derive from the *nous* organizing the whole, whereas in Aristotle's view, they stem from an "independent" form that cannot be definitively categorized as part of a united whole. In both cases, when parts exceed their limits, they transcend themselves, which ultimately leads to a diminished state of being. Limits are imposed on beings in a manner consistent with their intrinsic form or idea. Failing to adhere to these limits results in deficiency and privation. (We will see the significance of this later.)

In general, we do not find explicit or detailed arguments in Plato or Aristotle supporting the idea that teleology is *directed* toward man (anthropocentric teleology), even though one could argue that human rational capacity provides a unique access to the world (e.g., Rossi, p. 12).

This notion emerges more clearly in the 12th century with the concept of second-order natural teleology, expressed through the idea of *scala naturae* (the Great Chain of Being). In this framework, all parts are intimately interconnected within a functional whole, with God as the sustaining source of existence (*Summa Theologiae* I, q. 8, a. 1).

This reinterpretation of Aristotle is often characterized as anthropocentric, with its teleology being *external*. The ultimate goal is embodied in God's existence as the creator (*natura naturans*), which is distinct from the created things (*natura naturata*). The teleology is considered external because the form that shapes matter (God, or *natura naturans*) is not inherent to the substance (*natura naturata*) as a whole, as it is in Aristotle's view. Instead, the form is imposed from outside, a precursor to the idea that natural laws govern matter. In this hierarchy, beings serve a greater purpose, contributing to a higher end. It is anthropocentric in that humans are situated above beasts and below angels, with the rest of the natural world ordered to fulfill human objectives. However, this hierarchy is not merely functional; it is also *normative*. Proximity to God reflects not only rank but also moral elevation within the *scala naturae*. In this structure, everything has its designated place, meaning and moral status, leaving no room for randomness or chance.

This "deterministic" reinterpretation of Aristotelian teleology signifies the zenith of the *European whole-limiting cosmology*, prompting responses in subsequent philosophical thought.

Aquinas maintains that human beings are uniquely close to God due to their intellectual capacity (Glacken, 1967, 230). While God's essence remains unknowable, we can gain partial understanding through *reason*, as His creation is rationally ordered. Following Aristotle, Aquinas asserts that forms structure reality, enabling us to categorize and comprehend the world. He refers to these intelligible forms as *laws*, forming the basis of *Natural Law Theory*. These laws apply to all substances, i.e., there is a law for every (type of) substance. This alignment between the rational order of the world and human understanding is often referred to as Natural Theology, which has, in some respects, been seen as a precursor to modern natural science (Osler, 1994, 25 ff.). This suggests that the universe possesses an inherent order that is at least partially accessible to reason. Parts of the universe, such as humans, when guided by reason, navigate this order with a different clarity than those who do not.

Aquinas holds that true freedom lies in following reason (which aligns with divine order). Human beings are not mechanically determined but genuinely free to choose between good and evil (cf. the theodicy problem). However, our choices are most authentic and free when they are guided by reason toward the good (*Summa Theologiae I–II, q. 6–17*).

This freedom operates within divine providence rather than in opposition to it: God guides us in a manner that respects our rational nature instead of overriding it (*Summa Theologiae I, q. 83, a. 1–3*). Unlike later voluntarist thought, which posits a radical detachment of God's will from rational necessity, Aquinas maintains that divine order enables rather than suppresses human liberty. For some, this perfect order is reassuring; for others, it may evoke a sense of claustrophobia. Where did the contingency and randomness in life go?[8]

As Natural Theology gained traction, it also became more contentious. The tension arose from the concern that if reason can understand the structure of the world, it may impose limits on God's omnipotence. For example, does the rational order of the world reflect God's intellect in such a way that He cannot create contradictions (e.g., a married bachelor)? Furthermore, if a rational order is inherently good, does that imply the existence of a measure stick of goodness independent of God?[9]

One way to address this issue is to assert that God's primary faculty is not reason but will. God can, if He chooses, create a married bachelor. What God wills and what He does defines the good. There is no order preceding the order of God's creation. To claim that we can fully understand God is hubris. This perspective, particularly developed by William of Ockham, is known as voluntarism. (For a brief overview, see Haldane, 1989.) The concept of voluntarism leads to two significant ideas: first, that the order of the world can be radically contingent and, therefore, fundamentally incomprehensible; and second, that we can only know God or reality a posteriori—through *experience* (Osler, 1994, 30). Generally, these two ideas also characterize what later became known as nominalism, which is famously a precursor to empiricism.[10] Although many (Humean) empiricists would resist any metaphysical or ontological claims about the world, it can historically be said that empiricism represents a radical notion regarding parts and wholes. The implication is that wholes are merely constructions; what exists are only particulars, suggesting that we may be nothing more than parts that do not participate in an ordered whole.

Specifically, Hume argues that the external world consists of 'sensible parts.' These sensible 'parts' are properties, as Hume famously contends, there is no such thing as substance ('things'). Properties such as red, round, heavy and soft randomly coincide to form what we perceive as substances (Hume, 1978, 166). 'Anything may produce anything else' (Hume, 1978, 173), leading to Hume's characterization of the world as 'loose and separate' (Hume, 1978, 53).

The world is constructed from distinct, fully actualized properties (Hume, 1978, 634; Molnar, 2003, 181). The limits of what an entity can do are determined by its properties, but there is no necessary connection between being an entity and its properties; thus, there is no inherent essence of the (assembled) substance and, consequently, no real limits on what the entity can do either. This notion of a contingently ordered world is significant in this book and will be further explored throughout.

So, what does it mean to be part of a whole or to view the world through the eyes of a whole-limiting cosmology? Is it true that any concept of order necessitates a whole-limiting perspective? I believe not. A whole-limiting cosmology does not merely recognize order in the world; it also ascribes normative significance to that order—where any deviation is seen not only as disorderly but also as improper or morally wrong. Such cosmologies demand more than mere order; that order must also carry normative implications.

In contemporary contexts, these normative orders can be observed in the components that make up living organisms, but not in inanimate matter. For instance, we might say that the liver functions well for the benefit of the body as a whole. We also see this in the arts—such as plays, ballets and symphony orchestras—as well as in organizations like businesses and bureaucracies. However, these examples are from the human realm and do not represent broader cosmologies. While human-designed systems often impose limits on their components for functional or aesthetic coherence, the situation becomes more intricate when applied to natural systems. Do natural systems impose limits in a similarly normative fashion?

Current theories that may exemplify such whole-limiting systems include older ecosystem theories by Frederic Clements and Eugene Odum, Gaia theory (depending on interpretation) and more recently, concepts related to the strong anthropic principle (Goff, 2023). While these theories may suggest some form of normativity, they do not constitute axiological cosmologies or ontologies. The fact that theories suggesting whole-limiting order—such as certain interpretations of the Gaia hypothesis or the strong anthropic principle—are dismissed as fringe science more than indicates that contemporary scientific frameworks are generally resistant to cosmologies that impose inherent normative limits on nature.

Throughout this book, I will grapple with this tension. My argument is that the concept of order carries a weight that connects back to teleological and more broadly axiological cosmologies, which fosters skepticism toward 'order' itself. I contend that there is a certain order in nature—what I refer to as a 'historically evolved relational order'—but I will deny that order implies a whole-limiting order that imposes normative weight.

Before concluding this chapter, I will present a brief example of a whole-limiting cosmology that is not specifically linked to the European development of our current part-limiting cosmology.

In her paper *An Indigenous Philosophical Ecology: Situating the Human* (Rose, 2005), Debora Rose highlights key traits of the cosmology of the

Aboriginal communities in Timber Creek, Yarralin, Lingara, Pigeon Hole and Daguragu. Central to this discussion is the concept of The Country.[11] The Country is described as a semi-closed whole, comprising not only its people but also other creatures and nonliving parts. It 'comes into being through Dreaming creation,' and 'is itself the focus and source of Indigenous law and life practice' (Rose, 2005, 295). The Country possesses its own will, guiding its inhabitants on what to do, when and where (Rose, 2005, 299). To be a true inhabitant means being able to listen and interpret the signs provided by The Country (Rose, 2005, 297). Knowledge is not solely in the minds of the people; rather, it resides "in the Country." For example, "when the March flies bite, the crocodiles are laying their eggs, and when another type of fly arrives, the bush plums are ripe" (Rose, 2005). All creatures, including humans, are engaged in a collective effort to care for The Country. Everyone benefits from this shared responsibility, and if one fails to fulfill their role, they fail not only the other parts but also The Country itself: "You cannot pick all the berries, for then the Turkey will get none" (Rose, 2005, 301).

At some point in the year cycle, it is time to slash and burn (firestick farming). Our modern minds would see the firestick-burning as 'a disturbance of nature'; they see it as a necessary job which is critical to the maintenance of The Country and its parts. '*Burning The Country helps animals and people thrive; the benefit is* mutual' (Rose, 2005, 300, my emphasis).

This cosmology implies a kind of teleology that assigns normative significance to the parts. Rose articulates this idea as follows: 'The proposition has been that humans are part and product of country's self-organization. Neither the teleological end-point nor a random outcome, human beings are brought forth as essential and enmeshed parts of the country's life' (Rose, 2005, 302). Rose's seminal paper also reminds us that 'whole-limiting' should not be understood as the spatial organization of parts. Parts are organized in space and time. The country is a narrator, and the parts are part of a story. Where there is a whole that limits the parts, the story does not primarily link parts by chronological cause-and-effect sequences. Something happening now can be non-linearly "caused" (and limited) by future events or events in the past. The immediate causal mechanism explains how an event occurs, but not why it occupies its particular place within the unfolding pattern. (See e.g. Ingold T. (2016). Lines: A brief history. Routledge https://doi.org/10.4324/9781315625324). Famously, knowledge about how wholes temporally organize parts is prevalently preserved in songs and stories in Indigenous cultures. (See e.g. Struthers R. & Peden McAlpine C. (2005) Phenomenological research among Canadian and United States Indigenous populations: Oral tradition and quintessence of Time. In Qualitative health research., Vol. 15 No. 9. 1264-1276 DOI: 10.1177/1049732305281329

2.4 The rise of part-limiting hegemony

We have seen that whole-limiting cosmologies can be explicit and fixed to varying degrees. The Aristotelian view is a particularly interesting intermediate position. On one hand, parts possess an inner drive toward perfection;

exceeding their limits or the perfection prescribed for their kind leads to deficiency, privation and even monstrosity. On the other hand, Aristotle clearly states the independence of these beings (substances) (Phy. II. 8. 199b 15–16). Substances are primarily limited by their independent form, not by the whole. Whether this form participates in a united whole is not Aristotle's emphasis, at least not as it is for Aquinas. Medieval Aristotelianism evolved to increasingly highlight a more explicit and comprehensive whole-limiting perspective. Thus, alternative approaches only began to emerge when this view was fully articulated. As the concept of the 'whole' became clearer, disagreements arose over its exact nature, ultimately leading to voluntarism and what we now call nominalism.

As the whole-limiting view began to crumble from this point, a part of Aristotle that survived. *Intrinsicness.* Aristotle can be seen as the origin of what I term intrinsicalism (Lie, 2016, chapter 2.0–2.2), or the idiom of intrinsicality (see below). When form as a cause disappears (as in Aristotle's case) and we cannot know God's plan, *the question of what a real part is becomes pressing.*

The mystery is that the notion of intrinsicness persisted. Or rather, the question: What is x in itself? How can something be something in itself when the something has no inner telos? Which parts or ingredients can we regard as something in themselves? This question remains relevant in today's pursuit of objective knowledge, where we seek to understand what the object is 'in itself.' Even Hume, who rejected the reality of substances, implied a certain intrinsic feature in his ontology, which we now call quiddities—property-identity across possible worlds (Black, 2000). (A *minima sensibilia* would have to be something in itself.) In all instances, to ask 'what is x in itself?' is to inquire about the *intrinsic properties* of x, assuming such properties exist *independently of everything* else.

Most natural philosophers of the time adhered to some form of atomism. The real parts of the world are atoms (which originally means indivisible), and the world is constructed from these atoms. Thus, while the notion of "laws" (forms in Aristotelian terms) that Aquinas attributed to various (mid-sized) kinds remains, they no longer apply to the variety of kinds but to atoms. Kinds (form/idea) could be reduced to the laws governing these indivisible parts. Both atoms and mid-sized parts could then be viewed as passive. Consequently, independent external laws take on the active role of governing the parts.[12]

To accurately understand the world, we must reduce the laws from one level to the next, all the way down to the parts that 'really exist.' At the most basic level, we will find the proper *universal* laws, an ambition still reflected in the quest for a Theory of Everything. Preserving the laws while changing what they really apply to is also an argument for why we may understand these laws as deterministic. Laws applied to lower levels are justifiably rigid. Mid-sized entities that behave only somewhat regularly are inadequate if a proper explanation/description must provide absolute certainty and predictability. You need the necessity, which deterministic laws can provide. On this account, you would need descriptions that are exact, and those can be given only by what is quantifiable.

This rationale led Galileo Galilei, Descartes and John Locke to assert that only primary qualities (properties), such as size, weight and shape, are real, while secondary qualities, such as color, do not genuinely exist. Quantifiable properties exist within the object because it exists 'independently of everything else.' Color is merely a projection of the human mind. What exists within the object is the structure of its surface, reflecting light in a manner that makes it appear 'red.' Determining which properties exist independently of everything else and which depend on other things is crucial for defining objective knowledge. This distinction is also of utmost importance in this book, as one of the main claims will be that 'no property exists independently of everything else.' The notion that a real, objective property exists independently of everything is what I refer to as the *idiom of intrinsicality* (see also Sections 3.2.3 and 3.2.4).

The debate over the reality of parts is perhaps the most persistent and contentious problem in philosophy since the 16th century, which also turns the questions of ontology and metaphysics into epistemological issues.

The idiom of intrinsicality may explain why philosophers at this time began to reevaluate the relationship between ontology and epistemology. This shift from ontology to epistemology is often viewed as unavoidable. The premise is that we cannot say something about reality before understanding how to access it. We need to clearly differentiate between the parts called subject and object and find the right way to describe 'what contributes to what' in our understanding of the world. This shift becomes exceedingly important in Kant's *Kritik der reinen Vernunft*, which displays an impressive and obsessive attention to detail in describing the relationship between the 'inner' and the 'outer.' 'The question is how 'inside' and 'outside' relate to each other' (Bowie, 1990, 18; see also Section 2.10).

In the previous section, we noted that David Hume argues that the external world consists of 'sensible parts,' i.e., properties. 'Substances' are merely an idea projected by the mind. Consequently, we can define properties themselves as the primary constituents of the world. Moreover, we cannot assume anything about how these properties relate, as we cannot observe the relations themselves. Thus, there is no way to verify that snow *is* slippery. Snow (x) is a congregation of properties, but these properties might combine differently in the next moment, so to speak. This is why he claims that 'anything may produce anything else' and that the world is 'loose and separate' (Hume, 1978, 166 and 53). The world consists of (fully actualized) properties, so in Hume's view, there is a limit between properties in the sense that they differ, *but there is no limit to how they can combine*. Temporality is merely a series of discrete events (cf. perdurantism).

Grasping Hume's meaning is not immediately easy. We need to understand the view he opposes. Hume challenges the notion that there are necessary connections in the world, which could be an essentialist view. Essentialists (like John Locke, who was also confusingly an empiricist) broadly argue that substances (not primarily properties) exist and that these substances are bound to act in certain ways while being incapable of others. It is in their essence to do certain things, and this is how we identify them. They naturally fall under categories captured by our concepts (nominal essences) that define their capabilities, i.e., their limited potencies.[13] Importantly, however, their essence is defined not by their telos, *but by their composing 'minute parts'* (Real essence).

Moving away from the whole-limiting view, the general perspective appears to converge on a minimal idea of order: the world is not a gray soup but consists of distinct parts, with an inside that allows us to distinguish them from the outside. The nature of how these parts are related may be debated, but their independent existence is not. *First, we have parts; then we have relations*—most prominently causal relations. Some argue that these connections are random (neo-Humeans like David Lewis believe they are necessary within specific possible worlds), while others assert that laws necessitate them, even if the laws themselves are contingent (David Armstrong). Some maintain that the laws are not contingent, but only those operating on a fundamental level (Brian Ellis).[14] Regardless of the view on necessity, conducting science entails understanding the distinct parts and their causal relationships. This view constitutes the simple core set of assumptions in part-limiting cosmology. There is no overarching whole that regulates the relations among these parts.

Why is this significant? It is important because it appears that the whole-limiting view is replaced by a perspective that regards the parts as independent yet, at the same time, passive. They are passive in the sense that they can be rearranged 'in any way we choose.' If laws are viewed as (metaphysically) necessary, they are still only *necessary* at a fundamental level. Higher-level parts can be rearranged as long as they comply with these laws, which is quite liberal when considering parts we encounter at the human level (see Lie, 2016, 71). If the laws themselves are contingent, there may not even be a specific order to the world at all. We live in a world that could easily differ in myriad ways.

The passivist view developed during the transition from whole-limiting to part-limiting cosmologies is also linked to voluntarist perspectives that arose in response to Aquinas's interpretationist view. If there are no constraints on God's will, the products of His creative powers (*natura naturata*) cannot possess their own 'will.'

> One reason the mechanical philosophers were averse to the activity of matter, insofar as it is self-moving, was its potential to explain the world without invoking God or the supernatural. This danger could be mitigated if matter were viewed as naturally inert, producing effects only through mechanical impact; in this case, the source of motion in the world must lie outside the natural, material realm.
>
> (Osler, 178)

Thus, as God's agency gradually disappears from the acceptable ingredients in scientific ontology, there is an opportunity for humans to step into the same role if the idea of external agency acting on passively arranged matter persists. This is also why the relational ontology that I outline in Chapter 3 generally refers to the part-limiting ontologies as 'passivist ontologies.'

2.5 Part-limited freedom, ethics and politics

2.5.1 Freedom

In this book, I argue that the *ontological* turn described in the previous chapter has profoundly influenced how the novel and still vibrant idea of freedom was formed. As humans, we don't participate in a larger whole; the focus shifts to the interior/exterior issue of parts and a description of parts as discrete independent beings. Where do I start, and where do I end? When is this entity I call 'I' compromised, and when is it 'what it is'?[15]

This is where the issue of intrinsicness becomes important. What are the intrinsic and distinguishing properties that make me a human-being-part? The fundamental question of freedom shifts from 'how do I find my place within the larger whole' to 'who am I, and how do I differ from other parts?' Freedom becomes 'freedom from'—specifically, freedom from external impediments.

But what constitutes an external impediment? To determine whether I am impeded, I need to clarify where *I* end and where 'the external' begins (later expressed in the distinction between freedom of action and freedom of will). The race for clear distinctions between parts begins.[16]

In a part-limiting *cosmology*, freedom is reformulated as something contingent on the agency of the external world of parts, where these parts exert a coercive force on me as a part, rather than being something that constitutes or is part of me. This tenet is famously articulated by Thomas Hobbes:

> Liberty, or Freedom, signifieth (properly) the absence of opposition; (by opposition, I mean external impediments to motion;) and may be applied no less to irrational, and inanimate creatures, than to rational. For whatsoever is tied, or environed, as it cannot move, but within a certain space, which space is determined by the opposition of some external body, we say it hath not liberty to go further (…) A free man is he, that in those things, which by his strength and wit he is able to do, is not hindered to do what he has a will to do.
>
> (Hobbes, 1909/1651, 161)

Hobbes' view, later known as negative freedom, is debated, and a plethora of reformed views have been offered, but we would be misled if we overlooked the fundamental assertion that parts are fundamentally separate. Most debates on freedom begin with differing assumptions about the extent of this separation and *then only eventually so*: how are they related?

For *this* reason, it becomes essential for philosophers of the time to reconsider their accounts of the being of human parts and the being of nonhuman parts. It is from this vantage point that we need to understand human exceptionalism, which was previously asserted by God, who placed us below the angels in the hierarchy. If God no longer points his finger at us, there must be something inherent in humans that requires philosophical grounding. Various views have been proposed regarding how humans are distinct from other parts in terms of freedom, the ability to perceive other parts objectively, and the capacity to act ethically. These aspects are interconnected, especially when we consider how the other parts are framed and understood.

The emerging ideas about the human part are not the result of an isolated study of what it means to be human but are closely tied to how nonhuman parts are understood. This is simply because the new ontology of nonhuman parts changes our understanding of agency. John Locke, in his *An Essay Concerning Human Understanding* (1693/1995), may provide an example. In Chapter XXI, titled 'Of Power,' he addresses the ancient question of how to understand change. He begins by distinguishing between what he calls active and passive powers. Active powers involve the ability to *produce* change, while passive powers involve the ability to *receive* it (Locke, 1995, 165). Since all power 'relates to action, and there being only two sorts of actions (…) thinking and motion' (Locke, 1995), active power is most clearly demonstrated in thinking, according to Locke. Powers in nature (motions) can only be observed as *transmissions* of bare passion. Hence, 'The idea of liberty is the idea of a power in any agent to do or forbear any particular action' (Locke, 1995, 167). Furthermore, 'And power being the source from whence all action proceeds, the substance wherein these powers are, when they exert this power into act, are called causes' (Locke, 1995, 207). If agency comes from 'the outside,' the agent is 'under necessity' and therefore is not an agent but a patient. If the line of action starts *within* the potential agent itself, then the agent is the source of *production* of causes. This association between activity/power, causality and intrinsic nature is common in much philosophy of this period (see, e.g., Kant, 2002, 46).

An important claim in this book is that the mechanist, or what I prefer to call *the passivist* view, of nonhuman parts (described in the previous chapter) developed during this era of part-limiting perspectives, which foregrounds human exceptionalism. In other words, I claim that it is the *ontology of nature* that propels human exceptionalism.

I have noted that Hume doubted 'necessity in nature,' but this was not the prevailing view at the time. The dominant perspective was that nature is governed by laws, with God behind this law-governed universe (cf. the debate between Newton and Leibniz). Thus, if the parts of nature are moved by other parts in a deterministic way, human parts cannot act freely if we are part of nature. When the great chain of being is replaced by a cosmology where what we see today is a necessary result of a chain of causal events set in motion (by a Newtonian God or the Big Bang), external parts and the forces/laws that regulate their interaction emerge as the main threat to our freedom and our experienced human nature. Acting independently of this *chain of determined effects* becomes a template for understanding freedom.

The passivist view on nature may appear providential and useful in that it removes any normative obligation toward nonhuman parts. However, it creates a philosophical problem in defining humans. What are humans if not part of nature? While one might argue for Christian exceptionalism, this does not resolve the issue when philosophers of the time examined the details. If the other parts of nature are moved by deterministic forces, how can something that is supposedly outside this realm make a difference, and how can this realm not impact the exceptional (human) part? What kind of nature does this exceptional part possess? Surely it must have some *intrinsic* properties that make this possible.

The issue at hand is self-constitution. If part X is not constituted by something else (as is conversely the case with nonhuman parts), who or what constitutes the I or the self in a way that makes it independent from passivist nature—other than the self? From this viewpoint, the dualism between humans and nature seems inevitable, as it is already assumed. However, this dualism does not resolve the problem of self-constitution. The question still rings through: What makes humans an independent part?[17]

There are various answers to this question, with the most obvious tracing back to Aristotle. According to him, what makes substances independent is their goal-directedness. Although goal-directedness (final cause) no longer applies to parts of nature in a passivist ontology, it can still apply to humans.[18] While the other allohuman parts are viewed as acted upon by prior causes, humans can be an exception as they can act toward future goals. Kant famously argues this point. By referencing a specific human trait—rationality (Vernunft)—we can *transcend* what he calls the heteronomous kingdom of necessity. Rationality guides us toward future goals, whether ethical or instrumental. 'Conformity to the categorical imperative renders us autonomous, and conformity to the hypothetical imperative renders us efficacious' (Korsgaard, 2013, xii).

The idea in this book—that our basic notion of freedom emerges from a passivist view of *nature*—is particularly relevant in Kant's practical philosophy. Kant accepts this general passivist perspective, especially Newton's concept of inertia, which leads him to see nature as a *threat* to human freedom. External interference is viewed as problematic and must be countered by a faculty (rationality) that enables us to transcend antecedent causation through teleological action.

To sum up: In a whole-limiting view, different parts are ideally categorized by their contributions to the whole and valued accordingly. This was also the ramification of how "freedom" was a possible part of our lives. You are less free if you don't understand your role and how to act within your place. In a part-limiting view, where parts are passive and subjected to external and contingent laws, it makes no sense to 'follow nature' to achieve freedom. When 'the good' is removed from nature, freedom and ethics depart. Freedom is a prerequisite for ethical conduct, but ethical action (fulfilling the good) is not the source of freedom. The source of freedom lies in your ability to become independent from other parts. The general idea is that you can *increase* your freedom by reducing external constraints. This holds true, at least in terms of negative freedom. Thus, manipulating, defying and reshaping nature in ways that limit your ability to act as you wish is seen as necessary and positive for becoming human. People living 'in the chains of nature' are less human. They are savages (sometimes noble). In this sense, freedom and 'the good' align, but not in a direct moral sense.

From this vantage point, the European culture started on the journey toward emancipation not only from nature but also from religion, social norms, institutions, the state, our parents and our partners to finally become this one singular independent being. We call this progress, and freedom is outlined as the 'natural urge' to escape the limits imposed by the natural and social world.

2.5.2 Ethics

On a part-limiting view, ethics becomes theories about how we should relate to other parts seen from the *perspective* of those parts, rather than from the perspective of, for example, God. When we move away from the concept of a cosmological normative order, we face the challenge of determining which types of parts can be considered morally valuable. If the world is not a place where you fit into a preestablished order, but rather a space for independent parts that happen to congregate in certain ways, it seems that we are up against a buffet of parts that can be arranged in any way we choose. So, if this is our *starting* point, in which cases and for what reasons do I transgress the limits of other parts? That is: who and what should I respect and consider ethical counterparts? If we consult our ontological idea of what nature is, we find that there is no inherent value in nature itself. Nature is simply a series of causal events. As Freya Mathews puts it: 'They could not find any inspiration in nature regarding how to lead good lives since it was regarded as free of value and empty of meaning.' (Mathews, 2003, 4).

Baird Callicott expresses it this way:

> (I)n science, nature is conceived to be an objective and value-free system. From the scientific point of view, nature throughout, from atoms to galaxies, is an orderly, objective, axiologically neutral domain. Value is, as it were, projected onto natural objects or events by the subjective feelings of observers.
>
> (Callicott, 1989, 147)

What allows us to attribute value to certain entities based on their being? What constitutes real value? The part-limiting view presents only two options: we are either influenced by prior causes or by an inherent capacity for goal-directed behavior. Consequently, only goal-directed entities can be said to possess value that is independent of everything else (i.e., intrinsic value). There has been ongoing debate about what it means to act in this manner, with Kant being one of the strictest, who requires rational capacities as an *intrinsic* property to give rise to something having value (e.g., Muraca, 2011, 377). Today, after decades of environmental philosophy challenging Kant's stringent criteria, some accounts attribute goal-directedness to entities (including higher-order parts such as ecosystems) that exhibit some form of homeostatic activity (Lewontin, 2000).[19]

One might wonder whether the push for a strict exclusion criterion arises from rational arguments or our concept of freedom. For those already included, granting equal normative status to animals, other races, women, fetuses, rivers or ecosystems has always posed a threat to freedom, as new inclusions extend the external constraints that the already included must respect. It becomes a zero-sum game. My freedom is jeopardized by changes in the ethical status of other parts.

2.5.3 Politics

Another frontier where part-limiting cosmology had an inevitable transformative impact is in politics and political philosophy, specifically regarding the relationship between individuals and their shared societal reality.

How should we organize our society if the basic unit consists of discrete (intrinsically valuable) parts? My individual freedom is one aspect, while my freedom as a *citizen* is another. When hierarchical social systems are revealed as constructs imposed by those in power, it becomes clear that such systems are *chosen* and not derived from a natural order.[20] If we cannot avoid living together (cf. Rousseau), the best option appears to be establishing a social whole-limiting order that is consciously *chosen* and thus constructed/artificial.[21] The debate on freedom as liberty is, therefore, a discussion on the best *chosen* whole-limiting order.

The rise of (social) contract theory is usually attributed to Thomas Hobbes (although Rousseau was the first to coin and popularize the term), where he argues that our natural societal state is to combat one another. Hobbes' conclusion *follows from* the idea of individual natural rights (our birthright): If humans possess intrinsic natural rights, every individual is on equal footing. There is no aristocracy where the higher ideals of 'being human' are inherited. Our starting point is thus a 'war of all against all,' according to Hobbes.[22] Hierarchies are artificial (though they can be chosen, as in Hobbes' view). Our best rational choice would be to enter a contract where the state guarantees to protect my basic rights to life and health in exchange for giving up my right to do whatever I want against other humans. In societies accepting natural(ized) social hierarchies supported by a whole-limiting cosmology, there would be no need for the type of social contract proposed by Hobbes and later by Locke and Rousseau (the most influential philosophers of contract theory). The order, although locally different, is already in place. You belong somewhere. It is somewhat important to note that the 'state of nature' is not a historical reality; it is a thought experiment visualizing the situation if 'everyone had equal rights,' assuming that previous political orders in the past were merely another (albeit inferior) way to address the issue of equal birthrights. Solutions that ignored equal birthrights had to be justified by referencing a natural whole-limiting order (sometimes referred to as a sacred order), such as when a king was anointed by the church during the Early Modern Period in Europe. Contract theory is founded on an ontological notion of birthright, and the only way to organize a political system is to make it clear that any system at any time is and must be chosen.

This means that any proposed contract moving toward an independent whole-limiting order that is not chosen is inherently unjustifiable. Thus, liberals rightly argue that liberal democracy is not merely one among many possible political orders but the one that should be regarded as paramount. Despite the philosophical challenges of defining the essence of the individual part, the liberal perspective (which views individuals as the basic unit) holds the advantage. Any argument for subordinating to a (natural non-choosen) whole-limiting order cannot defend itself against deconstruction as an artificial construct. Appeals to something beyond the individual and its rights for a higher cause are suspect, especially when justified by referencing a nonchosen order (e.g., Daston 2019, 60). Thus, communism and fascism are questionable candidates for acceptance as chosen whole-limiting orders.[23] Additionally, any organized unit above the individual that 'takes on its own life' will be seen as a regression to a whole-limiting order, which in theory would require justifications entirely different from the liberal framework. Propositions that take humans to be basically social, dependent and relationally constituted, must hold on to 'a core of independence' to fall down on the right side of history.

As we discuss 'the reality of parts' and what follows from our assumptions, we may of course ask why we cannot deconstruct the individual similarly, i.e., 'individuals are just vehicles for their genes,' 'genes are just complex molecules,' etc. These are answers that science tends to provide. The fact is that we are not listening to science here. The reason is complex, and I cannot provide a full explanation here, except to say that politics as a tradition concerns how we should organize the parts we call humans. The part in a part-limiting view is the individual human being, which, according to liberalism, is the only part capable of exercising freedom and, therefore, the proper bearer of rights.

To recognize part-limiting views, we need to compare them to whole-limiting views and understand that part-limiting views develop as an *alternative* to whole-limiting views. The defining trait of a part-limiting view is that we are constrained by other parts, and their existence is viewed as a threat to freedom and *human* existence. This leads to a strong drive for transcendence. To transgress, rearrange or resist the influence of other parts represents a fundamental drive for those within a part-limiting cosmology. I have attempted to show that, despite potential hybrids between the two, the part-limiting view has consistently evolved toward its own distinctive features. The argument is therefore that if we want to find the origin of our modern-day unlimited behavior, we must return to our part-limiting cosmology. This perspective directly challenges accounts, arguments or intuitions that consider our unlimited behavior an expression of 'inner human nature.' Hence, the idea that humans are doomed to expand beyond their limits and that this is our predicament. I would rather say that this predicament reflects a historically contingent feature that we should attribute to the development of part-limiting cosmology.

2.6 Developments into our time

The part-limiting cosmology is a cosmology in the sense that it overrides. The race to the bottom to find the basic fundamental particle is still an ongoing obsession in science, *despite* the fact that we now know that fundamental particles are merely disturbances in a field (e.g., Higgs field or electromagnetic field) as measured by our devices. For the past 300 years, philosophers have struggled to define our inner nature, yet we cling to the assumption of a subject clearly separated from an object, which can be encountered at its boundaries. We still adhere to a billiard-ball concept of causal chains between discrete parts and discuss free will and voluntary action as if they hinge on whether our actions can deviate from these determined causal chains. Our ability and trust in predictable manipulations of nature rely on our capacity to separate and identify all involved parts. I am not suggesting these assumptions have gone unchallenged; rather, they have been seriously questioned and criticized, but this criticism has been overshadowed by the cosmology. A serious alternative has yet to be established.

In all these cases, we depend on specific and real limits. Fuzzy boundaries are unwelcome. When determining the limits of nature, we expect nothing less. We are obsessed with limits in a part-limiting cosmology, yet, as proposed in Section 1.2, we do not fully accept them.

This conundrum, I contend, arises from the part-limiting view discussed in the previous chapters.

At this point, I have two thoughts. First, I have no doubt that we live our lives within a part-limiting cosmology. However, we cannot see this unless we consider a third alternative: a relation-limiting cosmology. Furthermore, I am convinced that we cannot reach a better understanding of limits in nature unless we recognize that we actually live our lives within the outlook of a part-limiting cosmology.

Second, while the part-limiting view has consistently developed its own defining features, it has also evolved in the opposite direction. Beyond what has been said about the part-limiting framework, it is difficult to track down why the alternatives failed to gain traction. For instance, rationalism could have developed or at least preserved the Platonic idea of 'nous permeating everything,' but it was transformed into a faculty of the human mind by the Enlightenment movement, which needed "nous" to support individual freedom and the ability to distinguish, sort and conclude from clear premises. Romanticism, which can be seen as a reaction to the part-limiting view, was eventually summoned under the term idealism, only to become caricatured by what became known as 'analytical philosophy' in the mid-20th century (cf. Moore, 1903).

Thus, instead of following lines of thought that gained traction, I will limit myself to a particular strain of thought that came out of Immanuel Kant's *Critique of Judgment*, which inspired Friedrich Schelling to explore a different take on limits, distinct from the general view of the part-limiting trunk described here. Schelling thinks of limits as a "productive force" as opposed to a more spatial view where limits are seen as limits of parts.

It should be made clear to the reader that the naive idea of parts as isolated units, or humans as isolated subjects who can see, control and define the external world as isolated objects, was problematized from the very beginning. David Hume addressed this in the 'conclusion of this book' in his 'Treatise.' One could even argue that Kant's *Critique of Pure Reason* is a long rejection of the *naive* part-limiting view, as he attempts to establish how *things* are already interconnected and how we are bound to certain preestablished features of the world from which we cannot escape. As everything was torn into pieces, the effort to restore some kind of "order" began. However, this 'restoration' started from a new cosmological vantage point. The basic tenet of parts as the primary components of the world remained unchallenged. The goal was to demonstrate that these parts were not truly separate and that their connections were not as contingent as professed. The human self relies on others to achieve self-consciousness (Fichte and Hegel). Humans depend on each other to function on a societal level (Rousseau and republicanism). Parts *are* causally connected in ways that are not contingent (Kant). We do not perceive the world merely in terms of basic discrete sensory data (Kant).

Kant famously denied the possibility of knowing the world as it is in itself (*ding an sich*) and instead argued that we *can* know that our understanding of the world requires certain (transcendental) faculties of the mind to be experienced at all. There is a rational order to the world, but this order is attributed to the subject. We do not participate in the organization but are instead the organizing part (at least epistemologically).

For those who view(ed) this as a problem, the third critique of Kant (*Critique of Judgment*) may serve as a stepping stone that builds further steps out of this dilemma. In part two, he discusses the 'reality of the whole' in what he calls organized nature (biology). His general view (in the first critique) is that 'physical nature' should be explained through (mechanical) efficient causation (Kant, 2017, § 61). The world consists of independent parts that are subsequently *connected by* causation. This is why, I contend, he must anticipate that the 'wholeness' of the part is provided by the transcendental forms and categories in the transcendental subject, itself being a part.

The third critique allows for an alternative to the part-limiting assumption for organisms (organized nature), which he believes must be explained teleologically. This has consequences for how the parts are viewed:

> Now, the first requisite of a thing, considered as a natural end, is that its parts, both as to their existence and their form, are only possible by their relation to the whole (…) Its parts must in their collective unity reciprocally produce one another alike as to form and combination, and thus by their own causality produce a whole (…).
>
> (Kant, 2017, 201)

Epistemologically, this means that we cannot understand any (biological) part as distinct and separate in the first place. Every part must be understood in relation to other parts and in relation to the whole. The emphasis is not on the parts and how they build the world (bottom-up), but on the organization of the parts.

As the notion of (Aristotelian) teleology had been discredited by the time Kant wrote this, his reformulation of teleological explanations (in biology) profoundly impacted his readers and created a starting point for thinkers who disagreed with the idea of nature as passive and causally determined. While one can interpret Kant's third critique as a critique of his first critique (where his mechanistic view of nature and the dualism between humans and nature are emphasized), Kant himself did not fully do so. He regarded the notion of (biological) teleology as a *regulative* idea of reflective judgment (rather than constitutive judgment). Teleological explanations should guide our approach (epistemologically), but we should be cautious about ontological claims. Biological phenomena should be explained *as if* they are teleological, while non-biological phenomena should receive mechanical explanations (not explained *as* if). It is only in rational beings that we may find *real* telos and real purposiveness (cf. Kant's second critique).

Hegel (1770–1831) and Schelling (1775–1854) are the most significant early successors of Kant. Schelling develops Kant's view of our biological reality into an ontology of nature in general (Gare 46, 50 ff.; Schelling, 2004, 198). Hegel represents a different developmental line, emphasizing humans and human freedom. The reciprocity that Kant points to in his third critique is mainly deployed in his criticism of what I would basically call a part-limiting account of human self-consciousness and human freedom.[24] Here, I will focus on Schelling.

Schelling's critique of Kant's first critique is partly based on Fichte's views.[25] The possibility of "objective" knowledge presupposes a subject that recognizes its own existence and is aware of its activity. The fact that we are self-conscious is an intuition that cannot be further argued. Kant argued that the (scientific) subject was constituted by transcendental forms and categories, but Fichte contended that the (scientific) subject requires a more primordial experience of 'the other' through feeling (not perception or cognition) to become a transcendental subject (ego)—a subject capable of understanding and making judgments about the natural world. This understanding emerges through action; by interacting with the world of others (humans and other parts), the self feels that its (natal) freedom is limited by others. It is through *action* that the sensible world is constituted as objects, and only through reflection do we develop concepts of these objects (Gare, 2011, 37). Fichte argues that a basic proto-freedom must be in place for self-consciousness to develop. This "freedom" is a precondition for recognizing 'that which restricts' (i.e., other parts) and seeing oneself as different. Self-consciousness does not simply occur; to *know* the world presupposes the experienced re-cognition of the other as different. 'The other' is not merely an opposing part to an already self-constituted part, but a part that facilitates this self-constitution. The self is already relationally dependent on the other parts in order to be a self at all.

Schelling agrees with Fichte's criticism of Kant but makes an additional leap, aligning himself with a more whole-limiting view. He questions the givenness of the "object-side" in both perspectives. What we relate to in our subject-object relation is neither an atomistic manifold of sensations (synthesized by the transcendental ego) nor mere resistance from unrelated parts. His idea is that self-consciousness is a product of "self-limiting" nature.

Hölderlin argued that mutual recognition between the subject and the object presupposes a *whole* of which they are *parts* (Gare, 2011, 38). Hölderlin called this 'being.' Schelling referred to it as the absolute, which is seen as the condition and source for beings to have existence at all. The Absolute is itself unconditioned and prior to all opposition, while the being of the parts is conditioned (restricted) by other parts (Gare, 2011, 39).

What characterizes the Absolute is activity (viz. Anaximander). It is not only the subject and self-consciousness that are constituted by 'activity' (as Fichte emphasized). 'Activity' permeates all being, and through this unconditioned activity, the whole is *organized* (though it fails to be fully expressed in the parts). 'Nature is an ever-becoming product' (Schelling, 2004, 28). Thus, Schelling argues that nature is a historical product, denying both eternal laws and a goal-prescribing God (external teleology) (Gare, 2011, 52). Activity is both process and product (*natura naturans* and *natura naturata*) in one (Schelling, 2004, 22; Gare, 2002, 34). Process builds. *The organization that occurs is a process, but it is not the process that needs explanation, but its stable products (natura naturata)* (Gare, 2011, 48).

We are not up against things that *eventually* undergo a process of change. In a thing-oriented ontology, one seeks to explain how change and process are possible. Schelling, who develops a process-oriented ontology, must explain stability—the individuated permanence of things. He does this by extending Fichte's idea of restriction (which constitutes the Fichtean self) to all of creation. That is, through the limitation of different activities, X becomes recognizable as X. Activity is as much expansive as it is restrictive, and anything stable remains so through continuous (processual) restriction by other parts (Gare, 2011, 53). What we typically recognize as discrete and confined parts is an "assemblance" held together by opposing and reciprocal activity of the other parts. A stable being, like a rock or a human, is not first a thing that is eventually restricted, but rather a result of a constitutive process generated from restricting activity. Thus, to try to understand the agency of X, one must also understand the agency of the other parts in communion (Gare, 2011, 52).[26] When agency exists, it occurs within a mutually constituted "whole" (Schelling, 2004, 24). There is no activity in isolation. 'The two actions restrict themselves reciprocally, through interaction, to the mutual effect' (Schelling, 2004, 25). One cannot find the originating nature solely within the part.

By acknowledging the entirety of nature (not just humans or living beings) as active, it follows pretty straightforwardly that limitations must occur. A passive being that does nothing cannot be limited. Examples abound: a population of primula spreads up a hill until it encounters a limiting "force" (a temperature gradient, another competitive species, herbivores, etc.). Water fills a depression in the ground until it overflows the edge, establishing the boundaries of a lake. A certain molecular structure holds together until the temperature reaches a specific threshold.

From a human perspective, we may view these limits as opposing parts that restrict my activity as a part, or we may (to some extent) re-cognize them as factors that make our existence and actions possible.

An example from biology might clarify how this disagreement manifests in science. As a competing theory to the early geneticists of the 1950s (operating within the framework of what was termed the Weismann barrier), Conrad Waddington proposed that the development of an organism is not only generated by 'the (intrinsic) genetic blueprint' of the genome but also by restrictions in the *epigenetic landscape*, which, according to Waddington, channels how the genes operate within the cell (including when and how genes are expressed) (Waddington, 1957).[27] 'The channeling' represents restrictions *that make it possible* for the genes to function as they do (Noble, 2017, 168–169). While the specific activities of the genes may determine the amino acid sequence, this

> (I)s not sufficient to specify the phenotype. That also depends on the pattern formed by the relative quantities of the different proteins expressed. That pattern is not specified by the DNA sequence. It is *specified* by the cells, tissues, and organs of the body through the control they exert on the expression levels of different proteins.
>
> (Noble, 2017, 174)

Waddington's view aligns more closely with Schelling's general ontology, where properties and activity are neither produced by an inner essence nor by external governance, but rather by an ongoing activity that constitutes the parts through their mutual restriction.

At this point, I will not delve into Schelling's ideas about the relationship between the whole and the parts, or what he would term the relationship between the absolute and the finite, as this was a recurring struggle throughout his scholarship.[28]

It is undoubtedly challenging to pinpoint how Schelling's concepts of parts and wholes have influenced the establishment and development of the part-limiting view. His ideas represent a critique of the part-limiting perspective. Parts are not independent entities but rather products of a mutual arrangement of parts. Other parts, in a fundamental sense, are not a threat to my freedom; they constitute me and make my freedom possible.

However, it is difficult to argue that Schelling represents a revival of a whole-limiting cosmology. As the Christian God recedes in European erudite cultures, the only way to sustain the notion of a whole, given this historical context, is in terms of a *theoretical* "entity." From Descartes through Spinoza, Kant, Schelling and Hegel, the whole emerges as the *absolute* "negation" of parts. It is by virtue of the absolute that the parts exist as parts. In Kant, this absolute realm represents the ultimate pursuit of understanding the world (metaphysics), yet it remains unknowable. The whole signifies a boundary (*schranke*) beyond which what lies (our part-limited world) can never be comprehended in itself (Howard, 2021, 67). Schelling's attempt to unify these two realms is, therefore, often interpreted as a futile effort to restore (old-school) "metaphysics" or an endeavor to articulate the unspeakable. Either way, due to the general 'profanation of the whole,' the whole, as a concrete reality organizing the parts, seems to fade away. At most, it maintains a reminiscent presence in Heidegger's distinction between being and beings (the ontological difference) and perhaps even more so in his later works, where the 'difficulty of expression' becomes increasingly pertinent.

2.7 Limits *in* nature

The following sections of this chapter are still about examining the preliminary assumptions we make when discussing limits in nature. Thus far I have proposed that the way we think about parts and wholes constitutes one of these assumptions. What this preliminary assumption entails is the claim that we are more or less unable to provide an account that does not (naively) take one of the two as a *given* framework for answering the question 'what are limits in nature.' I suspect that the most likely naïve answer today would rely exclusively on the part-limiting view, where different part-limiting ontologies (Humean, essentialist, reductionist, etc.) compete to offer the best explanation.

The title of this book, *Limits in Nature*, presents several challenges. An obvious issue is defining what we mean by 'nature' and what we imply by stating that 'limits are in nature.' In the previous sections, I have attempted to provide a general overview of how limits have been viewed in the philosophical tradition. I have not delved into the details of different ontologies of nature, only discussing them to the extent necessary to support the two main views of how "limits" have generally been understood. I have also refrained from offering a specific perspective on how I believe we *should* define 'limits.' In Chapter 3, I will propose a third alternative to the part- and whole-limiting views, which will also clarify what I mean by 'nature.' What is lacking at this point is an assessment of what kind of philosophical questions that must be clarified when we want to challenge the current understanding of limits and freedom.

To initiate this discussion, we can start by examining what are sometimes referred to as reality principles. What kind of (explicit or implicit) principle do we follow when we assert that something is real? In the tradition of dispositional ontology, where I come from, it is common to distinguish between two main views.

One account takes something to be real when it is actual or manifested. If we can touch it, observe it, spray it, etc., it is real. This position is often called actualism and is associated with empiricism; Aristotle referred to it as megarianism. According to this view, limits are manifest—such as a fence, a membrane in our body or the surface of the sea—and are more closely tied to space than to time (to the extent we can distinguish between the two). This view often raises the question of whether these limits truly exist in nature or are merely constructs of our minds. A fence may serve as a limit for some but not for others, reflecting the old nominalist objection. Thus, only to the extent that we can assert such limits exist, we can also claim they exist *in* nature.

The other tradition holds that something is real if it possesses the ability or *capacity to do something*. This is often referred to as the 'Eleatic Stranger Principle,' which dates back to Plato. In the *Sophist*, the Eleatic Stranger states:

> I am saying that a thing really is if it has any capacity at all, either by nature to do something to something else, or to have even the smallest thing done to it by even the most trivial thing, even if it only happens once. I'll take it as a definition that those which are amount to nothing other than capacity.
>
> (Plato, *Sophist*, 242e)

'To be is to do or have the capacity to do.' From this principle, a quite different view on limits occurs. Limits are defining what beings are able to do. To the extent that we can say that such limits exist, we may also say that they are *in* nature, and not only 'in our heads.' On this account, we are not looking at the fence or the membrane as something real to investigate whether it is a limit or not; we are asking whether the fence or a membrane can and cannot do something.

As the reader may anticipate, there is a clear conflict between these views, as the actualist would reject potentiality as a form of reality (so-called *de re* potentiality), and defining "doing" is not straightforward either, as illustrated by Hume's famous analysis of causation. I will not spend time on arguing for the Eleatic view here, as it is thoroughly covered in Chapter 3, but I will state that the Eleatic view is the preferred perspective for our ongoing inquiry into limits in nature.

This decision should not surprise the reader, as it is already implied in the discussions of the part- and whole-limiting cosmologies. Other parts represent limits to what I can do, while the whole represents a limit to what I can (successfully) achieve. If we accept this view, it becomes crucial to ask where 'doing is done.' What is doing and what is not doing the things going on in the world. Here we return to the ontology of nature and, consequently, to what we mean by the term nature. Which parts are doing what? Is the agency located in the parts or in the whole? In what sense is it found in the whole, and in what sense is it found in the parts?

2.8 Where does the agency (and limitation) lie?

This is undoubtedly a complex and expansive question, and I can only offer guidance in certain directions to help the reader explore various options.

First, we can revisit whole-limiting cosmologies. In principle, a whole-limiting cosmology requires that the whole eventually assumes an agential role. It can influence the parts to pursue certain behaviors, as seen in Aristotle's view, where parts that fail to conform are deemed viceful or 'fallen' in the Christian tradition, or even monstrous. The whole can also compel certain behaviors from the parts, rendering them entirely passive, but where the whole (God) is principally enigmatic. This would be a theological universe where there are no restrictions on what God can do (cf. voluntarism). The doing is done by God and the parts place no restriction on this doing. Where the universe is unified by nous (Plato), every part would act toward a 'rational' order of the universe as a whole. In the case of Schelling, the whole strives for unity, with differentiated (nonsymmetric) activities creating and sustaining the parts through mutual limitation. The parts are constrained by one another and by their aspiration toward the whole (the absolute); they do not have predefined places within the whole (pace Aristotle). Instead, they come into existence as *something* from a 'grid-locked frozen relationship' among the parts striving toward the absolute. In this context, it is unclear to me where agency should specifically be attributed (see Section 3.2.4).

In part-limiting perspectives, it would be parts or partial elements/powers that do the job.

If we grant agency to the parts, we identify five options:

a Parts with essences possess properties that enable them to act independently of other parts (Ellis, 2002). They are primarily limited by their essence. For example, an electron behaves in specific ways due to its negative charge; without that charge, it ceases to be an electron. This applies similarly to giraffes, birch trees, herring, etc. Within causal frameworks (*causa efficiense*), a part's behavior is determined when acted upon. (Water will always boil when heated to 100°C). 'Essence-parts' can also be constrained by other parts in ways that obscure or mask their essence, preventing them from acting according to their true nature. A typical example (for illustration) would be humans who perceive themselves as having an essential core, with other parts either obscuring or facilitating their 'true nature.' This idea parallels the scientific experiment, which aims to isolate disturbances from other parts to reveal a part's essential nature.

b The prevalent modern interpretation of essence-parts is not Aristotelian (teleological) essentialism but Lockean essentialism. The Lockean view is that X has its essence not in virtue of its telos, but in virtue of its minute parts. (Locke 1690/1995, 212). This is how essentialism and reductionism combine. Here, agency is attributed to lower-level parts, which necessitate the observable (nominal) behavior. Often, this leads us to sub-atomic levels to identify parts with real (fixed) agency, suggesting that causality dictates behavior at higher levels. Consequently, higher-level parts, which humans interact with, are primarily limited by their constituent parts. This model explains part-activity as a *result* of both their essence and external influences. For instance, an organism's behavior is shaped by its genes and environment, itself lacking intrinsic agency.

c Alternatively, we might completely detach the part from its agency, asserting that it is determined by external forces (laws), meaning its behavior depends on the laws or forces of the universe. Few would argue today that these laws are metaphysically necessary; the common perspective is that they are contingent and could have been different (which is, in any case, to say that they are external to the part, as previously argued). This view is often held by neo-Humeans like David Lewis, who argues that while certain laws govern behavior of the parts in *this* world, they may govern the parts differently in other possible worlds. This means that properties can do anything (except what is logically impossible), but the assertion allows one to sidestep metaphysical claims about the origin and nature of laws. (If laws are deemed metaphysically necessary, their origin needs explanation.)

In option b, the laws are internal to the being, while in option c, they are external. In both instances, it is the laws (forces) that drive behavior.

d The Neo-Humean perspective does not necessitate a defense of radical contingency as Hume did. There are laws and some kind of order. Hume, on the other hand, posited that the world comprises properties that contingently converge to form "parts." These properties are separate, fully actualized and lack further potency to be limited by anything else (Hume, 1978, 166). 'Anything may produce anything else' (Hume, 1978, 173). Hume's modal dualist view suggests that what is not necessary *must be* contingent, leading him to assert that all relations are contingent. In this scenario, there is no agency at play, as everything real (properties) is fully actualized. What we call things or substances is just aggregates of properties which can be dissolved and rearranged without restriction (Mumford, 2004, 23).

What is common to c and d is that there are few reasons to talk about an 'interconnected world.' What we see as connections, for instance, that water quenches my thirst, that the Earth orbits around the sun or that salmon spawn in rivers, are just random events. Everything could have been different. Parts are separated in the first place and the only way they could be connected is by causal necessity. If that cannot be shown, they must be contingently connected. I will refer to this as the 'Humean view.'

e Hume's perspective was allegedly what awoke Kant from his dogmatic slumber. By adopting Hume's notion of causality, where parts are initially separate, Kant attributed their relatedness to the human mind. The limitations of the world (as described by science) thus become the limits of our understanding. What parts can and cannot do is ultimately constrained by our a priori transcendental faculties (categories and pure forms of intuition). What a part can do 'in itself' remains beyond our comprehension. By biting this bullet, Kant reached his goal of defending a causal view of parts, allowing us to attribute agency (in terms of necessary connections) to them in both scientific and everyday contexts.

These are the views I critique in this book. Accepting these perspectives, whether individually or in combination (see Lie, 2016, 70–71), leads to the notion that the only limits that exist are those that necessitate the behavior of the parts. Limits that do not impose such constraints are contingent (according to Hume) and, if posited, are likely products of our minds. If we embrace reductionism—an underlying theme in all these views—such limits (if they exist at all) would only include constants like the speed of light (299,792,458 m/s) and absolute zero (−273.15°C), among others.

2.9 Limits to whom?

Limits are limits to what X can do. The definition implies an X. A part. Unless we go by the idea that only subatomic parts are truly real, it would be misleading to assert that there are fixed limits in nature.[29] The speed limit of light, for example, is not a 'limit' per se; it is the maximum speed at which photons can travel. The limit on how much oxygen I can absorb into my blood at a given moment is a limit that applies to me, not to my cells, my neighbor or an elephant. When we discuss limits in nature, we are not referring to a uniform phenomenon; these limits vary based on the specific being in question. If limits *in* nature are constraints on what X can do, we must include X in our description, as far as we think that X is a part of nature.

The perception of limits as a uniform phenomenon is tied to our definitions. Within the current part-limiting cosmology, where I am one part among many, we tend to differentiate between the internal limits of the agent and the external limits of the world (see Sections 2.4). This framework suggests that, regarding what I can do, there are both internal limits within me and external limits outside of me. The limits in the external world are what we typically refer to as limits in nature or society. From this perspective, limits in nature are unrelated to X, i.e., this understanding of limits implies that parts are separate and self-contained (cf. intrinsicness).

In the previous section, I explored where various philosophical traditions locate agency: is it within the part, outside the part or is it both? I believe it is both, but not in the conventional sense of 'both,' where contributions are clear-cut and temporally distinct. What I mean by this will be duly investigated in the rest of this book. For now, it suffices to state that limits are intrinsically relational. From this perspective, limits are neither located within X (the agent) nor in the external world of parts. What parts can do is never achieved in isolation; it is always in conjunction with other parts.

This is entailed in the very nature of limits. Where something is limited by something else, there is also necessarily a relation occurring from the limit. As a matter of identity, what limits X from doing f constitutes the being of X (where 'being is doing') in the sense that it defines what X is not. This is also a basic tenet in Hegel's dialectics, such as the dialectic between the master and the slave. The more the master sees himself as a negation of the slave, the more he becomes dependent on the slave.[30]

If we accept this latter perspective, we need to reconsider how we view limits in nature. Limits are not external to X, nor are they internal to X. For instance, the uptake of oxygen in the bloodstream depends on the concentration of oxygen in the air. The oxygen available in the air limits how much can be absorbed into the blood, which is further influenced by the volume of red blood cells in the bloodstream. The gravitational pull on my body arises from the interaction between my body and the planet I am on. If something is defined by its actions (as posited by the Eleatic Stranger's reality principle), we should also recognize that relational limits are constitutive of who I am, as similarly proposed by Schelling.

At this point, we can assert the following about limits: they define what X can do, but they exist neither within X nor outside of X. Instead, they manifest instantaneously as mutual bonds between X and its partners. Therefore, we must carefully differentiate this view from one that perceives limits as something that *occurs in-between* parts, as that would reinforce the idea that parts exist prior to the limit.

2.10 On defining limits and being a part

Defining nature or the limits in nature is a performative action, as we position ourselves outside what we attempt to define. What we perform rests on an assumption deeply rooted in philosophical tradition, perhaps most clearly expressed in Descartes' famous dualism between subject and object. The radical subject/object distinction reflects the ontological idea that we are separate from what we define. What we mean by this might become more apparent if we try to define our own culture, of which we are more undeniably a part. To define something means we can delineate where it 'begins and ends' both temporally and spatially, which is quite challenging if you—yourself—cannot meaningfully distinguish yourself from what you are trying to define. In this scenario, defining your own culture becomes as much a cultural expression as an attempt to articulate that culture. You cannot define your own culture from an external perspective. In other words, the very act of defining your culture from an external perspective expresses the culture to which you belong. Thus, the effort to define nature and its limits appears to contradict a relational view of limits, whereas the classical part-limiting perspective (where limits are defined as either internal or external to 'X') is more coherent. The very attempt to define limits in nature therefore seems to presume a world of discrete entities constrained by other discrete entities, which, in turn, confirms a clear-cut part-limiting view.

As we have observed, this issue is part of a broader philosophical problem that has been recognized since Schelling. Schelling challenged the idea that nature can be fully captured by conceptual thought, arguing instead that it is a dynamic, self-organizing reality that transcends purely rational explanation. His contemporary, Georg Wilhelm Friedrich Hegel, viewed history as a dialectical process leading toward Absolute Knowing, where reality and thought merge 'behind our back.' One can also interpret phenomenology, particularly the Heideggerian version, as an explicit attempt to 'describe the world from within.'

To challenge the part-limiting view, we face at least two problems. First, as discussed in the previous chapter, a relational view of limits compels us to describe limits in a fundamentally different way. Second, we must reconsider our epistemological position. Where is the subject? If it is already a relational part of what it is trying to describe or define—an ingredient, as Graham Harman calls it—we find ourselves in a vicious circle (see Harman, 2007, p. 197 for the term 'ingredient').

With our relational view of limits, it seems we have placed ourselves in a corner of philosophy dominated by significant thinkers whose ideas we must explore to advance our understanding.

However, my approach in this book differs from the way this tradition frames questions and answers. I do not believe we are trapped in a vicious circle. This trap arises from the part-limiting view, and we need to closely examine *the idiom of intrinsicness* to trace the origin of this trap.

When we describe a being or entity (X), we outline its properties. An intrinsic property is one that X possesses 'independently of everything else.' This assertion, that beings have intrinsic properties which contribute to an accurate description of what X is, embodies the idea of (the possibility of) an objective description. To describe something from a neutral position is to articulate the intrinsic properties of X (assuming any X truly has intrinsic properties). So, how can the subject describe the object in an independent manner? By detailing its intrinsic properties. This current framing of the epistemological situation has posed a well-known problem for philosophers attempting to establish a part-limiting cosmology. With this premise, *we need to figure out our epistemic position before making claims about the world itself* (ontology). How can we access the 'thing in itself' if we, as subjects, possess certain inevitable faculties (such as language, prejudice, transcendental features and theories, as already discussed by Francis Bacon, who calls these idols)? They found themselves caught in what I term the 'epistemological dogma,' which G. Harman refers to as the 'philosophy of access' (see Lie, 2016, chapter 1.3 and 2021, 30, and Harman, 2013).

Following the epistemological dogma, one would find any attempt at doing metaphysics and ontology as naïve. In Kant's terms, such an attempt would entail the ambition to describe or understand the 'thing in itself.' But what about the dogma itself? Is the assertion that 'we cannot do metaphysics or ontology before we know how to access the world' as innocent as it seems? I would claim no! To state that something does have intrinsic properties is not ontologically innocent. On a relational ontology, for instance, there is no such thing as intrinsic properties. All properties are relational. Hence, on a relational view, we are not trying to access properties that are independent of everything else, which also means 'properties that are independent of the observer or the subject.'

The dogma, however would render this position as relativism, fortifying the realist position as a position where 'in itself' means 'independently of everything else.' This leaves us at a place where the dogma and its ontological foundation are immutable. Until this ontology is acknowledged, 'the relationality of nature' will be perceived merely as context and background conditions attributed to the subject's position—giving rise to the social constructionist perspective. The result is that we are left with a more or less naïve realism underpinning science and a social constructionist stance that acts as a proxy for a critical view of this very definition of realism (Lie, 2025).

My strongest argument against delving deeper into the philosophies of Schelling, Hegel, Heidegger, Merleau-Ponty and others is that they all operated within the epistemological dogma. None explicitly argued that the problem lies in the concept of intrinsicness, although Schelling came close. Had they done so, they would have proposed what I suggest here: we must adopt a naïve perspective and offer an alternative ontology to escape the vicious circle. We cannot (purportedly) state anything about the reality of the world without a reliable epistemology, yet a trustworthy epistemology cannot exist without ontology. In this context, I would advocate for a relational ontology over intrinsicness, describe it as accurately as possible, and then—*eventually*—position the subject inside this ontology.[31] This is what I aim to accomplish in Chapter 3 (Sections 3.2.4): Before describing this ontology, it is challenging to convey what I believe is possible in our effort to understand limits from within.

Even though we aspire to achieve a comprehensive understanding of limits in nature, we can only hope for insights that recognize ourselves as an ingredient from the outset. Whether we cross a limit or not depends on numerous other parts, both sentient and inanimate. However, this hope suggests that we may be able to theorize about what limits in nature truly are. We may also be able to describe how limits manifest in our everyday interactions (Chapter 4).

2.11 'It's not possible': The modality of limits

Another important aspect to consider is what philosophers say about the notion of 'possibility.' If something is deemed impossible, we typically say that we have 'reached a limit.' However, if it is possible to transgress limits, we must examine whether possibility or potentiality 'is a real thing.' If a certain limit is necessary, it cannot be transgressed; if a limit is contingent, it is merely contingent and not a true limit. If necessity and contingency are all there is to limits, there would be no limits in nature which is also possible to transgress. Limits in nature that are transgressed would not be limits. We have previously touched upon this, particularly in Section 2.8, when reviewing the 'Humean view on necessity and contingency. Some philosophers argue that potentiality does not exist, claiming that what occurs in the world is *either* necessary or contingent (random) (see Anjum and Mumford, 2011, 2017, for a further assessment of the modal dualist account).

Two important features are relevant to the modality issue at hand. When we discuss modality in philosophy, we refer to 'modes that are not the case right now,' typically expressed through modal verbs such as can, must and should. More systematically, these modal verbs correspond to traditional modal forms: necessity, possibility and contingency. If it is necessary that a certain pill cure a headache, it will do so both now and in the future. If it is possible that the pill will relieve your headache, it may or may not work right now. If it is contingent, the curing of your headache would be just a happy chance. This is referred to as modal force, with necessity possessing the strongest modal force. In this context, modality describes the mode of something, its way of being—related to time and space and beyond what exists right here and now. Therefore, when we say, 'you must do this,' we imply that in any similar situation, this would be the correct course of action, while 'should' carries lesser modal force. The stronger the modal force, the more limitations it entails. If something can never be the case, it cannot be the case right now either, because necessity entails that it cannot be otherwise (Aristotle, *Metaphysics Delta V* (iv)). If a photon cannot ever exceed the speed of light, it cannot do so at this moment either.

The central question here is what 'possible' means. Are possibilities merely "mere possibilities"? When we state that it is not necessary for Roger Federer to win the game, does that imply that anything is possible regarding betting on the game's outcome, except for what is logically impossible (e.g., Federer both winning and not winning)? Isn't there something about the world itself that makes it likely, but not necessary that Federer wins the game?[32]

Not according to modal dualism. Empiricism emphasizes observation as the gateway to knowledge, and we cannot observe the past or the future. So, what can we infer about the future based on the present? In any case, we cannot positively *observe* the potentiality or power of something right now either (Hume, 1978, 32). But can we know something about 'the here and now' based on the modality of X? Hume argued that to *know* that something will happen right now or in the future, we need to know that X causes Y by necessity. The challenge is that our understanding of the world is limited by what we can ascertain through observation. Hume famously contended that we cannot observe 'necessity.' His conclusion is that everything that occurs must therefore be contingent (Hume, 1978, 165). If it's not necessary, it is contingent! 'As objects must be conjoined or not … tis impossible to admit to any medium betwixt chance and the absolute necessity' (Hume, 1978, 171. See also 161–162 and 77). This is Hume's well-known argument against the existence of laws, but, importantly, he also posits that if there are any ordered or regular relations in nature, they must be necessary. If there is any order, then necessity follows. (For Hume and his contemporaries, connection and order imply that if something is connected, it is through causation, which is why he focused on necessity in his analysis of causation.) Hume's modal dualist view has had a significant influence,[33] and until recently, few accounts have argued for *de re* potentiality, or real potentiality (see Anjum & Mumford, 2017, chapter 2; Lie, 2016, chapters 4 and 5).

Philosophers and scientists have often operated under the assumption that statistics tell a story about connections in nature that are less than necessary and more than contingent, but this has not been grounded in a robust *ontological* theory. The argument is that statistics (probability) can ultimately be reduced to contingency or necessity (Anjum & Mumford 2017, 51 ff.).[34] We will not delve into this discussion here. The decisive point is that the modal dualist perspective supports a very strict notion of limits. While this may seem like a purely philosophical debate, we often encounter what is perceived as a truism in science and technology: 'Previously, we considered this impossible. Now, scientists have discovered a way to bypass the limits.' Limits that do not conform to necessity are viewed as contingent, and therefore not real limits. What seemed to be a limit turns out to be something that *could have been otherwise*. This is interpreted (from a modal dualist perspective) as 'what appeared to be a real limit in nature turned out to be contingent and thus a social construction.' Going with the modal dualist perspective, *we inherently disregard limits of nature that might be possible to surpass, but which are nevertheless a limit of nature*. Thus, we tend to include only those that can never be transgressed.

The social constructionist view on limits does not begin with the epistemological argument that we cannot truly know what nature is; rather, it originates from a *ontological perspective* on nature and how elements within it are connected: If we don't believe in everything we claim about laws and *necessary connections* in (e.g., are ceteris paribus laws and 'created in vitro'), then everything is contingent. Hence, a social constructionist accepts modal dualism and argues that if something could have been differently connected, we cannot have *realist* theories about any order of nature. Thus, any theory about how the world is connected would be socially constructed and discussing this subject matter (ontology) without acknowledgment of this is naïve (Hacking, 1999, 6).

This presents a challenge for the account I aim to establish in this book, which focuses on limits that are real, despite not being fixed or absolute (necessary). We are interested in limits that *can be* transgressed but are still genuine. In both our daily and scientific discourse, we assume there can be limits to how many elephants can inhabit an area, how much we can eat or how much sugar we should have in our blood over time, etc. While we accept this kind of 'limit-talk,' we do not seem to have a clear consensus on whether such limits are real. I *can* have more sugar in my blood than I should, so where is the (real) limit? Generally, most of us view the scientific and technological pursuit as an effort 'to chase nature's limits'. We are pushing against the limits of nature, repeatedly discovering that what was once thought to be a limit is not a limit after all. Does this imply that the limit was an illusion? It is difficult to determine what people mean when they say, 'we transgressed a limit we previously thought was impossible,' but they can certainly draw support from modal dualism if they wish to also argue that the limit was contingent and ultimately a social construction.

The alternative would be to argue that any limit is conditional (but not contingent). In this book, this means that they are relational (not *context-dependent*, which would be an epistemological phrasing). A limit refers to noncontingent *assemblage* of different dispositional partners that constitute a limit. Limits found in physics are also relational but appear fixed because of their rigid interconnectedness. This view will be further explored in Section 3.2.3.

Notes

1 In philosophy the study of part-whole relations is sometimes referred to as mereology. I will not engage in this field, however, because it traditionally leans toward logic and analytical philosophy, focusing on problems that arise *when parts are assumed to be* fundamental. For instance, the nature of parthood is often not closely tied to the kind of whole being considered, or whether there is a whole at all. That there is not any whole or "intrinsic relatedness" is taken for granted (see Harte, 2002, 13).

2 A cosmology is often understood as a theory about the universe as a whole and our place within it, suggesting that a cosmology must acknowledge the existence of 'the universe as a (united) whole' (Cosmos originally means order as opposed to chaos. See also Toulmin, 1982). Using the term 'part-limiting cosmology' may therefore seem questionable. When I refer to 'cosmology' (instead of ontology, for example), I aim to emphasize a more comprehensive perspective on how we view ourselves and the world, rather than proposing a waterproof theory about the universe. I use the term 'cosmology' when I want to convey an implicit and explicit 'self-evident' outlook on how the world is connected *and our place in* it. At times, I find the term 'cosmology' too formal, and in such cases, I may choose alternatives like 'view,' 'order' or 'trunk.' The expression part/whole/relation-limiting trunk underscores that a particular view is a branch of the same tree.

3 Physicists today discuss quantum fields (which remain largely mysterious) that "collapse" into particles. When something moves through a Higgs field (which is supposedly everywhere), it collapses into bosons that glue onto "the something" and provide it with momentum. A germ cell might also serve as an analogy, albeit at the expense of depicting something that is not truly unlimited.

4 Cosmos not only possesses order but also a good order. What explains the order entails not only a description but also a prescription. This is what we call an axiological cosmology.

5 Aristotle is here expressly arguing against the materialists (i.e., atomists) of his time who would explain wholes as aggregates of the parts: 'The cause of this nonaccidental unity is precisely the form, which is the principle of the living being as a whole, and hence the cause of each of its parts' (Rossi, 2010, 5).

6 This question is referred to as 'the problem of the existence of a second order natural teleology' (G. Rossi, 2010, 2).

7 The term 'reason' may be somewhat misleading in this context. Reason is associated with discursive thinking (*dianoia*), which involves calculation, inference and step-by-step reasoning—literally "thinking through" (*διάνοια*) something. *Nous*, by contrast, is (purportedly) our ability to directly 'see' with the intellect—an immediate, intuitive grasp of truth. As we will see, Plato holds that the universe itself is permeated by *nous*, and we partake in this reality precisely because we possess *nous*. Put more simply for those with a basic familiarity with Plato: we perceive the forms/ideas with the 'eye of the mind.' (Mason, 2013, 203).

8 Aristotle acknowledges coincidence as a part of existence; see, for example, *Physics* 196, b33.

9 For an in-depth discussion on this topic, see Osler, 1994, chapter 1.

10 Nominalism is best known as an alternative to concept realism, or simply realism, which is the belief found in the works of Plato, Aristotle and Aquinas that natural kinds—such as human-ness, elk-ness and gold-ness—are real. Nominalism challenges the traditional notion that our concepts accurately mirror the (natural kind) order of the world, as articulated in Plato's *Phaedrus* (265e). It questions the certainty that the particulars of the world belong to specific categories or essences that align with our universal concepts. This challenge to essentialism is crucial. According to nominalists, concepts—representing ideas, forms or kinds—are merely mental constructs that help us navigate the world. Therefore, we cannot assume our concepts correspond to any underlying cosmic order, which would allow us to use deductive reasoning and inference to uncover this order. This distinction is vital: if I assume such ideas, forms or kinds exist, I can deduce particular truths from universal knowledge. For instance, understanding that this fish is a cod allows me to predict its behavior without needing to observe every individual cod. When Hume enters the discussion, he questions the assumptions underlying this reasoning, introducing what is known as 'the problem of induction.' Hume challenges the concept of laws and our ability to know that a given cause will necessarily produce a specific effect. Some readers may recognize this as part of the ongoing debate about essentialism. This debate is not resolved by the arrival of nominalism and later empiricism. Instead, these perspectives shift our understanding of what it means to do science. While realists believed they could derive truths about the world based on a universal understanding of ideas and forms—similar to the scholastics—nominalism instills doubt about our knowledge of the world, urging us to move out of the armchair and engage with empirical evidence. The Copernican Revolution further propelled this shift, undermining the traditional scholarly reliance on ancient ideas. (I will use "kinds" or "natural kinds" from here on to maintain clarity.)

11 The term *The Land* is sometimes used to describe relationships to human environments that encompass not only the physical landscape but also the spiritual, cultural and relational dimensions that connect people to a place (see Whitt et al., 2001).

12 See Mumford, 2004, 147. This is how Brian Ellis formulates this tenet:

> The forces were not, however, thought to change the elementary things themselves, but only how they move or are arranged, for the identities of the elementary things were considered to be independent of the forces that operated them, and the forces were always thought to be external to the objects on which they directly impinge. So, the picture was one of intrinsic rigid bodies being pushed or pulled around by the forces of nature acting on them.
>
> (Ellis, 2002, 63)

13 Locke believes that our concepts often fail to capture the essence of a substance (belonging to a sort) as we typically cannot discern what constitutes that essence, namely the 'minute parts.' He therefore distinguishes between nominal and real essence (Locke, 1690/1995, 212).

14 The names in parentheses are examples of more recent philosophers who hold these views.

15 The obvious example here is Descartes, who went to great lengths to establish the reality of the "I" through the human rational soul.

16 A sort of 'evidence' for this is that we today, after a longer detour, still struggle to explain the empirical evidence of a field that influences the parts (electromagnetic fields), that parts have no location (nonlocality) and the empirical evidence of the

roots (cells' ability at the tip of a root) having the ability to 'know' which direction to grow to find water.

17 Some readers will notice that I intentionally avoid mentioning Descartes here. I want the reader to consider this issue without the specific associations that Descartes' philosophy would introduce at this point. Descartes is often regarded as the most influential "modern" thinker, but is that due to his own ideas or just the simple fact that he was one of the first to address the unresolved questions of his time? Questions that still persist today.

18 Famously, Aristotle's four causes (final, material, formal and efficient) were reduced to two: material and, more importantly, efficient (*causa efficiens* in Latin). This reduction occurred over time. However, by the time Kant arrives on the scene, it is clear that nothing remains. In his third critique, he states:

> (W)e are told that all this is in the highest degree contingent if we simply look to the nexus effectivus in nature, and do not call in aid a special kind of causality, namely, that of ends (nexus finalis). This means that nature, regarded as mere mechanism, could have fashioned itself in a thousand other different ways without lighting precisely on the unity based on a principle like this, and that, accordingly, it is only outside the conception of nature, and not in it, that we may hope to find some shadow of ground a priori for that unity.
>
> (Kant, 2017, 188)

What Kant suggests here prepares the ground for his assertion that biological phenomena may be explained *as if* this realm could adhere to a teleological explanation (see Section 2.2).

19 I mention Lewontin here as an example of a scientist that goes into great detail to show that living creatures are more than just the result of genes + environment. This reference highlights the importance of both independence and goal-directedness and their interplay. I am not suggesting that this is the primary goal of Lewontin's work. What I am pointing to is that this is interesting and important to us because it is important to determine whether some parts *are active or passive*. It is also important to keep in mind that most of the arguments that assign nonhuman parts ethical status, particularly intrinsic value, within environmental ethics focus on goal-seeking properties (see, e.g., Pratt, 2000, Chapter 7).

20 Cf. the geocentric cosmology where all of God's creatures are placed in a hierarchical order.

21 The artificial or constructed nature refers to political theories from Antiquity and the Middle Ages, where the polis was seen as the natural "habitat" for human agency. This distinction matters because claiming that social and political order is natural leaves little room for liberty. If constructed, it implies that the system could have been different. Thus, political arguments appealing to nature are gradually viewed as illiberal.

22 Using the term natural rights in the context of Hobbes is somewhat misleading, as it was Locke who first consistently formulated this idea within the framework of a social contract.

23 Communism is not easily categorized. Marx's argument was a typical part-limiting stance: he claimed that a capitalist society, despite liberal assertions, is one where the whole has taken on its own course. To liberate the individual, we must unite in a revolution to combat the system exerting unacknowledged and pervasive downward influence on its parts. Whether his proposed replacement system would exemplify a chosen whole remains a contested question.

24 Although I don't think he overcomes the part-limiting cosmology.

25 I will mainly follow Gare, 2011 in my presentation of Schelling's view on limits.

26 Schelling believes that the creation of consciousness and self-consciousness is a historical process in which nature understands itself from within. The I that attempts

to understand the world is a part that seeks to comprehend the whole of which it is a historical product. To position the I outside the world (in a dualistic subject/object relation) would be to regard it as a stable, unconditioned part, which it is not. Understanding the world from within is necessary but only possible to a certain extent (Gare, 2011, 39, 52).

27 Waddington is often referred to as a biologist who challenged what Francis Crick called the Central Dogma in the early stages of genetics (Crick, 1958; see also Cobb, 2017).

28 The parts originate from the absolute (the whole) but fail to express the absolute, and this (potential) dualism haunts him. This topic is, of course, a recurring issue in philosophy, although some philosophers—such as the empirical positivists—deride it as 'mere metaphysics.'

In the *First Outline of a System of the Philosophy of Nature*, he argues that the whole is primordial to the parts (198) but denies a 'teleology of the whole.' He believes that the whole exists in the parts in terms of 'striving towards the whole,' but the whole itself does not strive for something. While the parts strive for some form of organized order, they fail to express the whole. Epistemologically, this means that one cannot understand the whole through understanding the parts, despite the fact that the parts are a result of the activity of the whole. The whole (the absolute) does not express itself through laws or specific organizing principles. The expression of the whole consists in the activity of the parts, which fail to fully express the whole.

29 When I refer to 'fixed limits,' I go along with the assumption that laws are lying, which is what Nancy Cartwright argues (Cartwright, 1983). I agree with Cartwright, but there is little sense in bringing this to the table at this instance.

30 For those who want to dig into Hegel's more abstract expositions on this matter, see *The Science of Logic* (Hegel, 2010), chapter 2, p. 98–99, in particular.

31 As argued in Lie, I believe this approach differs from Heidegger's perspective in *Being and Time*. Heidegger essentially accepted the epistemological dogma when he refrained from attempting to offer an alternative ontological description of nature and instead sought to characterize the dominant ontology of science as a consequence of a specific epistemological stance (present-at-hand). A more effective solution, as previously discussed, is to view this stance as a result of a particular ontology (passivism) and thus propose an alternative (relational) ontology. Others, such as Graham Harman, Karen Barad and Quentin Meillassoux, have partially proposed similar approaches (Lie, 2025).

32 I refrain from making the question of modality more complicated by introducing David Lewis's modal realism here (Lewis, 1986).

33 Especially in framing the debate. Instead of questioning the modal dualism proposed by Hume, opponents of his view have attempted to argue that there truly are necessary connections in nature. This began as early as with Immanuel Kant, who contended that necessity is a transcendental feature that makes thinking and experience possible (i.e., taking for granted that for causation to exist at all, there must be a necessary connection between cause and effect).

34 It is controversial whether the dispositional theory can account for this modality, but in my opinion, Anjum and Mumford make a strong case in "What Tends to Be" (2017).

References

Anjum R. and Mumford S. (2011). *Getting causes from Powers*. Oxford University Press.
Anjum, R., & Mumford, S. (2017). *What tends to be*. Routledge
Aquinas, T. (1989). *Summa Theologiæ*. Ed. T. McDermott. Ave Maria Press.

Aristotle. (1996). *Physics*. Trans. Robin Waterfield. Oxford University Press
Aristotle. (1998). *The metaphysics*. Trans. H. Lawson-Tancred. Penguin.
Black, R. (2000). Against quidditism. *Australasian Journal of Philosophy*, *78*(1), 87–104.
Bowie, A. (1990). *Aesthetics and subjectivity: From Kant to Nietzsche*. Manchester University Press.
Callicott, B. (1989). *In defense of the land ethic: Essays in environmental philosophy*. State University of New York Press.
Cartwright, N. (1983). *How the laws of physics lie*. Clarendon Press.
Cobb, M. (2017). 60 years ago, Francis Crick changed the logic of biology. *PLoS Biology*, *15*(9), e2003243. https://doi.org/10.1371/journal.pbio.2003243
Copleston, F. (2003). *A history of philosophy* (Vol. 1). Continuum Publishing Group.
Crick, F. H. C. (1958). On protein synthesis. *Symposium of the Society for Experimental Biology*, *12*, 138–163.
Daston, L. (2019). *Against nature*. MIT Press.
Ellis, B. (2002). *The philosophy of nature*. Acumen
Gare, A. (2002). The roots of postmodernism: Schelling, process philosophy, and post-structuralism. In C. Keller & A. Daniel (Eds.), *From process to difference: Between cosmological and poststructuralist postmodernisms* (pp. 34–50). SUNY Press.
Gare, A. (2011). From Kant to Schelling to process metaphysics: On the way to ecological civilization. *Cosmos and History*, *7*(2), 26–69.
Glacken, C. J. (1967). *Traces on the Rhodian shore: Nature and culture in western thought from ancient times to the end of the eighteenth century*. University of California Press.
Goff, P. (2023). *Why: The purpose of the universe*. Oxford University Press.
Hacking, I. (1999). *The social construction of what?* Harvard University Press.
Haldane, J. (1989). Voluntarism and realism in medieval ethics. *Journal of Medical Ethics*, *15*(1), 39–44. https://doi.org/10.1136/jme.15.1.39
Harman, G. (2007). On vicarious causation. *Collapse*, *2*, 187–199.
Harman, G. (2013). An outline of object-oriented philosophy. *Science Progress*, *96*(2), 187–199. https://doi.org/10.3184/003685013X13691199842803
Harte, V. (2002). *Plato on parts and wholes: The metaphysics of structure*. Oxford University Press.
Hegel, G. W. F. (2010). *The science of logic*. Trans. & Ed. Di G. Giovanni. Cambridge University Press. (Original work published 1812).
Hobbes, T. (1909). *Leviathan*. Oxford University Press. (Original work published 1651).
Howard, S. (2021). Kant on limits, boundaries, and the positive function of ideas. *European Journal of Philosophy*, *30*, 64–78. https://doi.org/10.1111/ejop.12652
Hume, D. (1978). *A treatise of human nature* (2nd ed.). Oxford University Press. (Original work published 1739).
Johansen, T. K. (2020). Plato's teleology. In J. K. McDonough (Ed.), *Teleology*. Oxford University Press.
Johnson, M. R. (2005). *Aristotle on teleology*. Oxford University Press.
Kant, I. (2002). *Critique of practical reason*. Trans W. S. Pluhar; Intro. S. Engstrom. Hackett Publishing Company. (Original work published 1788).
Kant, I. (2017). *Critique of judgment*. Trans. J. C. Meredith; Rev. N. Walker. Oxford University Press. (Original work published 1790).
Korsgaard, C. (2013). *Self-constitution: Agency, identity and integrity*. Oxford University Press
Lewis, D. (1986). *On the plurality of worlds*. Blackwell.
Lewontin, R. (2000). *The triple helix*. Harvard University Press.
Lie, S. A. N. (2016). *The philosophy of nature: Rethinking naturalness*. Routledge.
Lie, S. A. N. (2025). Making sense of 'description from within': An ontological challenge to (eco-) phenomenology. In M. T. Heneise, et al. (Eds.), *The Routledge handbook of Ecophenomenology*. Routledge.

Locke, J. (1995). *An essay concerning human understanding*. Prometheus Books. (Original work published 1690).
Mason, A. (2013). The nous doctrine in Plato's thought. *Apeiron, 46*(3), 201–228. https://doi.org/10.1515/apeiron-2012-0068
Mathews, F. (2003). *For love of the matter: A contemporary panpsychism*. SUNY Press.
Molnar, G. (2003). *Powers: A study in metaphysics*. Oxford University Press.
Moore, G. E. (1903). The refutation of idealism. *Mind, 12*(48), 433–453.
Mumford, S. (2004). *Laws in nature*. Routledge.
Muraca, B. (2011). The map of moral significance: A new axiological matrix for environmental ethics. *Environmental Values, 20*(3), 375–396. https://doi.org/10.3197/096327111X13077055166063
Noble, D. (2017). *Dance to the tune of life: Biological relativity*. Oxford University Press.
Osler, M. J. (1994). *Divine will and the mechanical philosophy*. Cambridge University Press.
Plato. (1993a). Phaedrus. In J. M. Cooper (Ed.), *Plato: Complete works*. Hackett Publishing.
Plato. (1993b). Sophist. In J. M. Cooper (Ed.), *Plato: Complete works*. Hackett Publishing.
Plato (1997). Phaedo, in *Plato complete works* (ed. J. Cooper). Hackett Publishing Company.
Pratt, V. (2000). *Environment and philosophy*. Routledge.
Rose, D. (2005). An indigenous philosophical ecology: Situating the human. *The Australian Journal of Anthropology, 16*(3), 294–305. https://doi.org/10.1111/j.1835-9310.2005.tb00312.x
Rossi, G. (2010). Aristotle on the rain, once again. *Dianoia, 56*, 91–123.
Schelling, F. W. J. (2004). *First outline of a system of the philosophy of nature*. Trans. K. R. Peterson. State University of New York Press. (Original work published 1799).
Toulmin, S. (1982). *The return to cosmology*. University of California Press.
Waddington, C. H. (1957). *The strategy of the genes*. Allen & Unwin.
Whitt, L., et al. (2001). Indigenous perspectives. In D. Jamieson (Ed.), *A companion to environmental philosophy* (pp. 3–19). Blackwell Publishing.

3 Limits in nature

Introduction

Biologists would typically say there is an approximate limit to the density of a population of ptarmigan in a given area. If the population density exceeds a certain level, intestinal parasites will spread rapidly, leading to a population decline. We rarely find reasons to challenge scientists when they conclude, "There is a limit to the number of ptarmigan in this population." However, when it comes to what humans can do, 'limits' become more complex, as we often find ways to overcome them. For instance, we might administer antiparasitic medication to the ptarmigan population to reduce parasite density. From a scientific perspective, both statements, 'there are limits' and 'there are no limits,' can be true. Can a better philosophical understanding of limits clarify this?

Perhaps. This is at least what this chapter aims to achieve. We have examined whole- and part-limiting views on limits. In this chapter, I will explore what a 'relation-limiting view' entails, delving deeper into the ontological understanding of limits. Ontology addresses fundamental questions in philosophy, such as worldviews or the nature of things. In this case, ontology concerns the question, 'What are limits from a relational perspective?'

There are many 'ontologies of relations' on offer, and there are many signs of relationalism being on the rise (see Section 3.2.4). In this book, I have chosen what I see as the most promising account. This account builds on what is often called dispositionalism, which is the ontological view that reality consists of bearers with powerful properties, or causal powers. However, dispositionalism is not a relational ontology per se, so we need to explore how we can develop this ontology into what I call relational realism (Lie, 2016, Chapter 4). From here, we can begin to frame the philosophical understanding of a relational view of limits in nature developed in this book.

3.1 What science says about limits in nature

If we aim to understand what kind of ultimate limits science is talking about, we need to ask whether there is some point at which there would be nothing beyond it. Such a point would signify that 'nature is completed.' This would be an end state representing an absolute limit that nature 'cannot surpass.'

DOI: 10.4324/9781003307372-4

As previously discussed, modern physics is characterized by its rejection of a teleological cause in nature. Nature does not strive toward an end state; it 'does not go anywhere,' meaning there will always be something beyond, a limit to transgress. But is this metaphysical identity of modern physics strictly accurate?

In contemporary physics, we observe that parts have a preferred "collective" end state, known as symmetry. According to thermodynamics, nature "seeks" the lowest possible energy state in a system, including the system as a whole. Where there is difference or asymmetry (such as in temperature differences), nature "seeks to equalize the difference" (in this case, leading to windy conditions on Earth). What we perceive as order is actually a state of differences, but the fundamental order of nature is, according to physics, "chaos": a maximally symmetrical state (maximal entropy) where there are no differences among the parts and where all energy (and matter) is equally distributed. In a state of equal energy potential, there is also inertia, where nothing occurs and the potential for action is zero.

When equally distributed parts combine to form "wholes," such as stones, umbrellas, birch trees, water and humans, some form of external energy is required to maintain their differentiation from their surroundings. Without this external energy, these different entities will revert to lower energy states. The umbrella, stone and so on, will eventually degrade into "less differentiated" smaller parts that ultimately decay into "atomic chaos." The differences we observe in nature are thus various types of *temporary* differences and limits. According to this perspective, the identity of X is not fixed, except for the beings we find at the subatomic level that are bound to chaos and cannot be broken down further. Therefore, everything is in transition toward symmetry. The only true limit in nature is the end state of total equilibrium, with no "further" end state and thus no limits beyond this.

When I say there is a limit to what I can do to a certain person or a tree, I view this unit as having an identity that, at a certain point in development, cannot be surpassed without becoming something else. This identity is not inherent; rather, its nature is to be in continual "transcendence" toward the ultimate goal of nature: complete equilibrium. Any limit we impose beyond this is "not real." Consequently, it would not be incorrect to describe the limits we are concerned about in environmental issues as socially constructed. They represent a temporary state of order, implying that this order can be considered chosen, at least to the extent that humans can change it. When we kill someone, destroy the Amazon or pollute the oceans, it may seem we are violating limits, but in truth, we are not. These limits are not real; they refer to a specific human-defined order that has not yet been challenged. From a physicist's viewpoint, this order might even be seen as one that goes 'against nature.'

From a biologist's perspective, the situation appears different. The phenomenon of life is often said to defy nature's tendency toward symmetry or the second law of thermodynamics.[1] By creating and sustaining particular complex structures, organisms generate more order, i.e., less entropy. While a non-living stone slowly decays into less organized, smaller parts, a living organism continuously maintains an internal order that resists the pull toward higher entropy from its environment (until death). For instance, animals can maintain a constant temperature, salinity and fluid balance, "regardless" of external conditions. Organisms adapt to their environment by utilizing available free energy sources from nutrients and the sun, either directly or indirectly. This "negating capacity" is regarded as a defining feature of the life phenomena. Erwin Schrödinger (1944) called this negentropy. The details of this concept may be debated, but the notion that the biosphere constitutes an order distinct from the entropic tendencies of physical systems is widely accepted. This is easily illustrated with simple examples: a wound heals, an animal thrown off balance by the wind regains stability, a plant in drought develops deeper roots, plant communities progress toward a certain climax state,[2] and the liver functions to remove ethanol from the bloodstream.

From this, we can conclude that there must be an "additional" scientific concept of order in nature. Philosophically, there is a basis for asserting such an order linked to the part/whole-limit discussion. Are we willing to claim that the wholes we observe in biological phenomena are true unities? While physics asserts that anything different is merely a preliminary state tending toward symmetry, biology argues that differentiated life forms are genuine unities *because* they 'limit themselves.' The ability to limit oneself from external influences is fundamental to the *unity* of the whole (i.e., organisms). In this context, the wholes found in biological phenomena should not be viewed as contingent aggregates awaiting further division into smaller parts down to a symmetrical state of indifference. If we accept what physics states about limits in nature, we must also recognize the unities that negate this tendency toward symmetry as real. In other words, they should not be dismissed as social constructions. A stone cannot die, but a self-organized living entity can. Death or non-existence is a real limit, not merely an intermediate state. Life is a limiting activity, and in this sense, we can infer that there are limits in nature at higher levels.

Acknowledging this limiting activity as real, the limits that humans should focus on are those that lead living systems toward lower entropy. To understand what this entails, we need a way to describe this order. In this book, I refer to this order as *the order of possible-N*, though at this stage, it remains a speculative idea requiring substantial philosophical grounding.

3.2 A dispositional view on limits: Introduction

The 'ontology of dispositions' or 'dispositionalism' is a recent development within the analytical philosophy camp, with roots tracing back to Aristotle (Cartwright, 1994; Ellis, 2001; Harré & Madden, 1975; Heil, 2004; Marmodoro, 2010; Martin, 2008; Mellor, 1974; Molnar, 2003; Mumford, 1998, 2004) Dispositionalism is often seen as a strong competitor to neo-Humeanism, which denies the existence of any worldly potentials or causal powers. Consequently, the debate on dispositions can appear quite embedded and 'scholastic' to those unfamiliar with the analytical tradition in philosophy.

In this introduction, I aim to present this ontology of nature to a broader audience, demonstrating that it can do more than challenge neo-Humeanism. I see dispositional ontology as an appropriate starting point for a perspective on nature that portrays it as fundamentally active (challenging *passivism*) and relational. To reach this point, many doors must be opened and closed, and the work required to fully justify the account would require a book of its own. This presentation here draws on my previous work (Lie, 2009, 2016).

Most relational ontologies developed recently have continental rather than analytical philosophical roots. The account presented here encompasses both traditions. Broadly speaking, analytical philosophers engaged in metaphysics and philosophy of science often neglect issues that extend beyond (analytical) philosophy, while continental philosophers typically find the analytical method and *lingua* unappealing and irrelevant. A strength of analytical approaches is the attention to the details of the argument. Although analytical philosophy has its blind spots, it adheres to an ethos of clarity and transparency. The language is less loaded and more amenable to scrutiny, which makes it easier to apply this ontology to the philosophy of science and how science understands nature. Many continental philosophers discuss science and its *role in society*, leaving the analytical philosophers to address the internal workings of science with some authority. I consider it crucial to find a way to critically examine the internal workings of science in relation to its reductionist and part-limiting tendencies to understand how this aligns with the role and external workings of science. This is why I prefer to use analytical philosophy to articulate such views. Furthermore, I believe that dispositional ontology serves as a superior starting point for a relational account of nature, effectively and convincingly challenging neo-Humeanism, essentialism, reductionism and social constructionism (Lie, 2016, Chapters 2 and 5).

In Section 2.9, I argued that limits define what x can do. To support this claim, I referred to the Eleatic Stranger principle, which is also the foundation for dispositional theories, which generally claim that something is real if it can do something.[3] I will now provide a broad introduction to this theory and illustrate how it can be extended to the issue of limits, depending on our interpretation of the theory's implications.

When we ascribe properties to something, we describe what we believe that thing is. For example, we say 'seagulls fly,' meaning that seagulls possess the property of being able to fly. What does it mean to be able to fly? A dispositionalist interprets this as the ability to perform a certain *behavior*, an ability that does not disappear when the seagull walks on the sandy beach. But what does it truly mean to say that a property is a specific behavioral activity of something? Why do dispositionalists emphasize behavior and activity? The answer is simply that they attribute the behavior to the being in question. A being that loses its property when not manifested is not genuinely or intrinsically active. A bird that loses its ability to fly whenever it doesn't fly must be given the property from something else whenever manifested. Humans "can" fly, but only if seated in an airplane. To assert that 'dispositions are real even when not currently manifested' is to claim that the property is located within the being in question. The power resides in the being itself. This is why dispositionalists argue that laws originate from dispositional properties. Laws do not describe an additional ingredient in the world that supplements the powerful dispositional properties.[4] This allows dispositionalists to describe the world as fundamentally and genuinely *dynamic*. Things possess potencies, or causal powers, that can be exercised (*energeia*) in various ways depending on the situation.

3.2.1 Are all properties dispositions?

Dis-positions? Is there really something in the world that exists but does not have a position? This is the dispositionalist claim. Dispositions are properties that can exist even though we cannot point to where the property resides. We all agree that seagulls can fly, but where is this property when a seagull walks on the sandy beach? Has the seagull lost this property whenever it is not exercised?

'Properties' play a significant role in our language. We use 'properties' to designate what something is. A dispositionalist will assert, as previously noted, that these ascriptions describe what something does or is able to do, but the range of what something is able to do is never fully manifested at the same time. Dispositionalists believe, as we have seen, that these unexercised properties still exist in the being in question. The ability to fly does not vanish when not exercised. It is not a power external to the seagull that comes into being whenever it flies; it is a *real* property of the seagull (see for instance Martin, 1994, Mellor, 1974).

Are all properties dispositions? Here we should note that participants in the debate on dispositions disagree. The so-called categoricalists (e.g., David Lewis, David Armstrong and W. V. Quine) argue that dispositions refer to *a way of talking about properties*. Only categorical properties truly exist (e.g., Armstrong, 2005, 315; Lewis, 1997, 149; Quine, 1974). Categorical properties are those that are continuously and fully actualized. They are properties that cannot do anything more than they do (Lie, 2016, Chapter 3.2). Categoricalists argue that these are the fundamental properties of everything to which any other type of property can be reduced, such as spin, charge and mass. Dispositions are merely "placeholders for the arrangement and interactions of small bodies and the laws that govern them" (Quine, 1974, p 10). This view aligns with materialism (or physicalism), although this does not imply that dispositionalists align themselves with non-materialist positions (see Lie, 2016, Chapter 3.2).

The dualist, on the other hand, maintains that there are two types of properties—dispositional and categorical, and insists that dispositions *cannot be reduced* to categorical properties. This perspective spurs the same familiar types of questions that dualist views raise elsewhere. For example: If they are different, how are they connected and so on. Mumford (1998, 191, ff.) and others distinguish between dualism and aspect-dualism or what Mumford calls neutral monism. We will see that I hold a neutral monist view, which I refer to as a 'revised neutral monist view' because I regard the dualist accounts (sometimes referred to as mixed views) as not genuinely departing from categoricalism (see Lie, 2016, Chapter 3.4 and section 3.2.3 in this book).

A third possible view is that categorical properties do not exist at all. This position is called dispositional (or power) monism, or pandispositionalism. While categoricalists claim that categorical properties are "passive," and that it is the *external* laws that do the work (in other words, categorical properties do not *do* anything, which is the main point of being categorical), the dispositional monist would then claim that such categorical properties do not exist. This follows more or less from the Eleatic principle that dispositionalists adhere to: *for something to be, it must do or be able to do something*. The dispositional monist thus argues that there are no external laws governing passive beings. "Laws" are intrinsic to the beings themselves, as Stephen Mumford (2004) illustrates with the book title 'Laws *in* Nature.' What we call laws are in fact derived from dispositions (Bird, 2007, 44; Mumford, 1998, Chapter 10; 2004 Chapter 10.6). This perspective resembles how Aristotle sees potencies, powers or dispositions: as internal to the thing in question.[5]

3.2.2 How can 'something potential' be real?

When the dispositionalist, through the Eleatic principle, claims that potencies are 'in the thing' as real existing properties, we call *de re*, or *worldly*, potentiality. It exists neither in our heads nor externally to the being in question. Our intuitions both support and contest this claim. On the one hand, it seems strange to assert that something that is, by definition, not manifested (dis-position) at the same time exists. How can something that is not "there" exist? On the other hand, it seems odd to say that the property disappears when it's not manifested. For example, the carpenter does not lose the skill to do carpentry the moment he sits down (Aristotle's example, *Metaphysics* 3, 1047a).

Humans (and organisms in general) act as if beings indeed have *de re* potentiality. We do things to make something manifest (i.e., fry the potatoes), and we take measures if we fear that a certain disposition might manifest (i.e., spill some of the boiling oil on our child). In science, we try to determine whether X has a certain disposition. To have knowledge in a field is to understand how certain things may or may not behave. Logically, X may behave in all kinds of ways except those that are logically inconsistent. But X does not in fact behave in all kinds of ways.[6] The ways in which X can behave are *limited*. X is 'more than it is,' yet also less than everything logically possible. The future is not entirely open; not anything can happen (contra the Humean world of pure contingency, where anything that is logically possible *could* happen for all we know).

So how do we limit the range of X's dispositions? Essentialists within the disposition camp restrict behavior to what they consider essential dispositions (e.g., Brian Ellis, 2002). Others ascribe real potentiality only to dispositions that are "truly" fixed (Alexander Bird, 2016). Neo-Humeans, like David Lewis (1986), attempt to reduce *de re* potentiality to logical possibility, spelled out in terms of possible worlds. Charlie Martin (2007, 31–32) argues that the dispositional range of X is extensive, though still narrower than the full scope of logical possibility. I will return to Stephen Mumford's and my own position in the next section, but I should mention already now that I distinguish between the range that X has naturally and its 'maximal range,' which forms the basis of many arguments in this book.

3.2.3 What are the dispositions had in virtue of?

Going along with the first intuition mentioned in the introduction to the section above, it seems reasonable to assume that dispositions must have some kind of "physical aspect" present in X. As we have seen, the categorical monist claims that these "physical properties" are all there is, and that these categorical properties are 'governed by laws.' Dualists argue that both are present, but that instead of laws, we have dispositions that inhere in the 'physical aspect,' if you will. Dispositions arise from categorical properties, but they cannot be reduced to such properties. I have argued elsewhere that most of these positions ultimately resemble categorical monism (see Lie, 2016, 104, esp. footnote 64). Dispositional monism, as we have seen, claims that categorical properties do not exist. They assert that dispositions arise from *other* dispositions (at lower levels) and/or relations (Mumford, 2004, 174). Mumford contends that dispositions become *fixed* through their relations to other dispositions (Mumford, 2004, 182 ff.; 2007, 148).

In the last section, I said that Stephen Mumford originally held a "dualist" view under the broader notion of 'neutral monism' (Mumford, 1998). This account entails a *level-view of* nature, where the categorical base serves as a foundation for higher-level dispositions, but where categorical properties and dispositions are not properties in their own right (Mumford, 1998, Chapter 9.9). The categorical property is categorical *in relation to the dispositional aspect on the level above* and vice versa. (Some readers may recognize a similarity to the previously mentioned *hyleo-morphean* view of Aristotle, where matter and form are mutually dependent). The main reason Mumford abandoned this view was his conviction that 'categorical properties don't do anything' and therefore cannot influence dispositions. If they did, they would be dispositions (Mumford, 2004, 150).

While I agree with Mumford's relational view, I also believe it is both possible and beneficial to retain the neutral monist view.[7] This may seem a bit awkward, but upon closer examination of what 'categorical properties' truly are, we may realize that what has been characterized as categorical properties in this debate could be understood as *manifestations of dispositions at a lower* level. By making this shift, we do not introduce an extra element of categorical properties into our ontology, and we avoid the burden of making sense of how categorical and dispositional properties are related (the dualism problem). Moreover, we can uphold the idea that dispositional properties are all there is (dispositional monism). Manifestations are already part of the dispositional framework, and we simply assert that manifestations are 'doing something.' This way, we achieve the best of both worlds. *Natura naturata* and *natura naturans* are one. Thus, H_2O (along with other manifesting partners) manifests water, which, when manifested, possesses further dispositions, such as being drinkable, enabling osmotic pressure and regulating electrolyte concentrations in our cells. Water is also known as a universal solvent, which is also a dispositional property (Lie, 2016, figure 4.3).

As we have seen, categoricalists argue that dispositions can be reduced to the categorical base (what I call the manifestation aspect or manifestation base), while dualists contend that they cannot. The dualist argument primarily seeks to justify this claim by asserting that dispositions are real. I believe this is insufficient and that the dualist, as mentioned above, ends up being a *de facto* categoricalist and reductionist. This is where relationalism plays a significant role. Water is not drinkable due to its molecular structure (the manifestation at a lower level) but rather because of other manifestation partners, like humans and other animals. The fluidity is not solely determined by its molecular structure, but also by the temperature surrounding the water. Dispositions are not fixed by laws or their molecular structure but by other relational partners. (Fixed meaning 'water *is* fluid'.) The nature of X—which dispositions (i.e., properties) it has—depends on which other bearers (with their dispositions) are present in the universe, but which dispositions that are in fact manifested depends on the local presence of other bearers (Lie, 2016, Chapter 4.5). The range of dispositions of X is therefore wider than we usually think, which is also why we tend to wrongfully believe that the bearers have essences. This also explains why we tend to think that we can predict how for instance water will behave (manifest) based solely on its molecular structure, and that the molecular structure *determines* the behavior of the bearer, while in reality, this regular behavior should be explained by the *regular* presence of other bearers. Thus, this version of relationalism provides strong reasons to reject reductionism in general. The dispositions of water are not entailed in the manifesting base (H_2O) of water.

The modified neutral monist view (Lie, 2016, Chapter 3.6), where the categorical base is the manifestation of dispositions at a lower level, combined with a relational account of dispositions, results in what I call *relational* realism (Lie, 2016, Chapter 4.6). This is the view I will refer to throughout this book. Dispositions arise from the manifestation aspect, combined with dispositional partners. You cannot have any dispositions where there is no manifestation (e.g., H_2O), and you cannot have these *specific* dispositions (e.g., drinkability) without the specific relational partners of the bearer (e.g., humans). The range of 'what X can do' is wide but is limited by its manifestation aspect and other dispositional partners.[8]

One further dimension is needed to fully articulate relational realism: *the historicity of dispositions*. Dispositions are, on the relational view that I promote, historically evolved.

Drinkability is clearly a historically evolved dispositional property—but arguably, this is true of all properties if we accept that the universe we inhabit today is a product of natural history. (The reader may compare this general perspective with Lee Smolin's concept of *temporal* naturalism, where he argues that the laws of nature are historically generated, in contrast to the prevailing view that they are timeless.)

The relational realist view that I have briefly outlined above relies on the recognition that nature is structured in levels. Atoms are composed of subatomic particles; molecules are formed from atoms; more complex entities, like DNA, are composed of specific molecular arrangements—and so on.

If we consider what most cosmologists believe about the history of our universe, we can briefly reconstruct the early stages of this development as follows. In the early stages of the universe—during the Great Inflation following the Big Bang—only quarks and gluons existed. However, tiny quantum fluctuations in an otherwise uniform energy field introduced slight irregularities in density. These were amplified by gravity, eventually leading to the formation of cosmic structures.

Quarks and gluons obviously exhibited the disposition to combine into protons and neutrons (what is known as the strong nuclear force), showing the further disposition to manifest as an atomic nucleus. This is where the disposition of the charge occurred. In the presence of electrons, the nuclei "became" positively charged, and in the presence of the nuclei, the electrons "became" negatively charged (known as electromagnetic force), allowing the formation (manifestation) of stable atoms. These atoms represent a new level of organization—a stable (durable) manifestation that could give rise to further dispositions occurring in the presence of partners capable of "touching" or interfacing with that level.

Throughout this historical development, bearers gain new dispositions. For instance, when water becomes drinkable due to the emergence of life, it takes on further dispositions, such as enabling cell *membranes* in "collaboration with" fatty acids (which are hydrophobic).

With the unfolding of natural history, more layers of relatively rigidified (continuous) manifestation architectures emerged, making it possible for further relationally constituted dispositions to manifest as processes, objects or events. It is an 'epiphytic' universe. Emergent phenomena are often dismissed or only accepted in special cases, such as the occurrence of life and consciousness. On this theory, emergence is the rule (see also Lie, 2016, 5.2.)

3.2.4 What relational accounts require

Relational ontologies are gaining popularity in both theoretical output and scholarly engagement. Movements such as new materialism, post-humanism, environmental humanities and enactivism (epistemology) mark key directions in these developments. Thinkers like Karen Barad, Graham Harman, Bruno Latour, James Ladyman, Tim Ingold, Anna Tsing, Donna Haraway, Jane Bennett, Gilbert Simondon and Félix Guattari are notable in this context, with many tracing intellectual lineages back to figures like Baruch Spinoza, G. W. F. Hegel,

Friedrich Schelling, Alfred North Whitehead, Henri Bergson, Martin Heidegger and Maurice Merleau-Ponty. Some have developed explicit ontologies, while others are more suggestive or implicit.[9]

I will not engage the reader in scrutinizing this field here. Instead, we will focus on how a relational account can provide a different understanding of limits, aiming to establish a convincing theoretical foundation for the viability of such an alternative. Others may use different relational accounts to explore how their preferred relationism connects to limits. In all instances, it is important to substantiate key requirements for developing a relational ontology that fits our purpose here.

Firstly, if we are interested in limits in *nature*, we need an explicitly and genuinely realist account. Thus, when discussing relational realism, we must demonstrate why relationalism is not relativism and show that what makes properties real is not their intrinsic nature. (I will return to the question of relativism later in the section below.) We have already seen how the manifesting aspect "steps in" to prevent a relapse into intrinsicalism for the *dispositional monist account* (which is considered the proper framework for a relational theory of dispositions). Intrinsicalism is the common denominator of part-limiting views that seek to distinguish between real and non-real properties of the parts. Objectivism is the idea that certain properties of the (external) object exist 'independently of everything else,' which is what we aim to access when trying to understand 'what the world is, in itself' (cf. the epistemological dogma discussed in Section 2.10).

Secondly, we must account for relationalism without implying that relata disappear into the relation. I have briefly shown how the manifestation aspect "picks out" the parts involved, making it epistemologically possible (though challenging) to localize these parts and identify which disposition belongs to which partner. At the same time, the relational nature of dispositions dissuades us from concluding that the disposition belongs to the bearer 'independently of everything else' (see Lie, 2016, Chapter 4.6).

Thirdly, our account must not be a reductionist one, where relationality is merely a feature of the Basic Level, ascribing realness to this level alone (see, e.g., Martin, 2007, Chapter 4–4). Some relational accounts, supported by physics, suggest that everything can be reduced to a foundational level where relationalism serves as the proper ontological account (i.e., structural realism: Ladyman & Ross, 2007). In such cases, any part may be combined with another as long as it aligns with the relational order at the basic level 'where the real stuff occurs' (see also Bird, 2016). If the only (real) limits in nature are found at the basic level, discussing limits in the environmental debate offers little value. What is proposed here is instead an anti-reductionist relational account: it is the very relationality at each level that prevents reducing the relational order on one level to the level below.

A more technical way to express this argument is to say that if something is real by virtue of what it does, it is evident that bearers on lower levels do not exhibit the behaviors we observe on higher levels (Lie, 2016, Chapter 4.5.2). Let's take an example. A popular way to talk about serotonin today is to say that "serotonin makes you happy."[10] However, this is not what 5-hydroxytryptamine (serotonin) actually does. Serotonin signals by binding to certain receptors. If we accept that the *identity* of something is determined by *what it does*, we must be precise in describing what it actually does. Thus, it's inaccurate to attribute happiness to serotonin. Happiness is not a disposition of serotonin, but of a person. Similarly, a *table* is not just a collection of atoms because tables serve different functions than atoms (pace Eddington's two tables). Practically, we cannot build a table by collecting atoms and assembling them. What defines a table is what it does, its dispositions. Similarly, *glass* is fragile, while molecules are not.[11]

As we are concerned with limits in this book, it is important to clarify which partner we are addressing. What limits our happiness is not the same as what limits serotonin signaling. A categorical base (manifestation aspect) does not determine the higher level; it restricts ("tracts") the higher-level dispositions (like being 'low on serotonin'), but these dispositions are also shaped by what the bearer encounters, including its interfacial partners (e.g., the death of your child). I may relieve a headache by changing the biochemical partners involved (and thus become less limited), but the pill does not remove the dispositional partners contributing to the organism's stress.

Fourthly, a proper relational account must accommodate that we are inevitably part of what we seek to understand when we explore and transgress limits through our manipulations. If we are part of a relationally constituted nature, it follows that limits are also relational and are experienced as such, albeit interpreted differently. Here, we seem to face only two options. The first is to claim that we are merely experiencing a certain 'slice of reality' or 'agential cut,' as Karen Barad puts it. This perspective is often linked to relativism, suggesting that limits are simply what we perceive and experience, thus fundamentally constructed. However, relativism is unacceptable for the realist, which may even drive the (classical) realist to reject relationalism. So, what is our option if we deny both positions? Our option is to describe the ontological situation we find ourselves in (see also Section 2.10). This is what I will attempt to do in the following sections.

3.2.5 Why relationalism is not relativism: Relational realism

First, we must reject the notion that dispositions are free-floating properties. Dispositions belong to what dispositionalists call *bearers*. It is not properties (dispositions) that constitute the bearer; rather, properties are ways the bearer behaves (is) (Martin, 2007, 44). Bearers are composed of bearers that come together to mutually manifest (on the lower level). Furthermore, as we have seen, bearers are not necessarily things or objects: manifestations can be (what we recognize as) continuous ("objects"), discontinuous (events) or processes. There is no conflict between object ontologies and process philosophies on the relational realist account. What we are observing is bearers acting in specific ways. In discussions of dispositions and relations, the bearer often gets sidelined.

Dispositionalists tend to focus only on the status of dispositions, "forgetting" the entity that actually bears them (e.g., Mumford, 2007). When we discuss an active, dynamic world and aim to reject passivism, we must not overlook the bearer.

Having emphasized the importance of the bearer, we can now address the relationship between ontology and epistemology. We may ask: what does it mean for something to be genuinely relational? What distinguishes "relationality" from real relationality? A seemingly relational view is one that interprets relations as primary causal relations, while acknowledging that a phenomenon (e.g., a headache) can have multiple causes. The idea would be to map these causes as individual contributions to the phenomenon through an experimental setup that isolates each cause to identify its unique contribution, which can then be added up. This form of relationalism is both fragmentational and compositional. It assumes that the partners (bearers) contribute *independently* of one another, allowing the researcher to map and synthesize the "complex" network of contributions. In such cases, there is recognition of multi-causal instances, but not of true relational behavior. Real relational behavior, in contrast, is reciprocal and mutual, meaning that what X does depends on what Y does.

Within the disposition camp, real reciprocity is problematic because one can only certify that X has a disposition based on its manifestation (this is called isomorphism). To say that wood has the disposition to catch fire implies that this (single) disposition will manifest under certain conditions (x, y and z). However, claiming that "a single disposition manifests" when x, y and z occur violates the fundamental idea of relationalism. If a *manifestation* is always mutual and reciprocal, we cannot directly observe the manifestation of a *single* disposition. In other words, there is a conflict between the requirements of a dispositional theory (isomorphism) and relationalism. Whether this issue pertains solely to relationalism depends on how you interpret "when x, y and z." In this case, I believe a relational interpretation is appropriate. "When x, y and z" clearly indicates that the manifestation occurs only when these other partners are present; thus, the manifestation must be a mutual co-production. Therefore, we need to abandon isomorphism as far as isomorphism requires epistemological confirmation.

I argue that this is not a problem for the dispositional theory, but rather for science (Lie, 2016, Chapter 4.4.1). Specifically, it poses a challenge for science to the extent that scientific endeavors must lead to clear, transparent and apodictic knowledge, which, according to relational realism, is highly problematic. Scientists do not face a neat set of separate building blocks that behave consistently and independently of one another. A genuine relational account suggests that these "building blocks" interact reciprocally. This brings us back to the significance of the bearer. It is the bearers that enter the situation, not the dispositions themselves. The bearer is flexible and possesses a wide range of dispositions, with its contributions depending on the other partners involved. This is what we mean by saying that X has a variety of dispositions.

The epistemic problem here is that you cannot know the contribution of one partner (bearer) before you understand the contributions of the other partners, since each bearer does not contribute with the same disposition independently of the others. For example, it is the different contribution from the surface (reflecting light) that causes the apple to change from yellow to green, not just the eye or the light; understanding this requires knowing the contributions of both the eye and the light. Is it the eye that contributes a different disposition, or is it the surface structure of the apple? The crux here is to understand that the contributions of different partners can change depending on the others involved (e.g., someone with color blindness). This can be further clarified by contrasting this account with the realism/relativism debate.

As previously noted, the core issue of the part-limiting epistemological regime is the problem of access (cf. the epistemological dogma). If we seek to access the intrinsic being of the bearer, the subject must remain outside what is called the object. The relational account presented here contends that this is impossible. We are already *in media res*, acting as relational partners participating in manifestations simply by observing or conducting experiments. According to the epistemological dogma, this position is untenable because it leads science toward relativism and social constructionism. Thus, if light behaves as particles when measured by humans, it suggests (at least according to certain interpretations) that humans *construct* the manifestation.

The alternative is to say that humans (including measurement instruments) are manifesting partners like any others. 'Light' does not have an essence. It has different dispositions that can manifest with or without human partners. In a specific experiment (double slit), light manifests in this way or that. With other partners, it may (or may not) manifest differently. The main point is that manifestations involving humans are as *real* as those occurring without them! *It is not constructed.* It is the *same "thing,"* 'light,' contributing differently due to *a different disposition*, resulting in a different mutual manifestation. There are no noumena: nothing beyond or behind what we observe. *However, what we observe is not the whole truth about the real potentialities of the partners involved.* When I see the color red, it is not an illusion just because the manifestation includes me as a partner. There are no primary (or secondary) qualities of color. *The physical structure reflecting light is as relational as the color*, and this goes all the way down.

As I previously mentioned, the task of science is not to reach objective truth (i.e., the intrinsic—independent properties of X) because there is no essence or intrinsic quality. The task of science is to dis-cover which partner contributes which disposition (at each level). Because dispositions are reciprocally related and manifestations are always mutual, this is quite challenging, but not impossible (as the case with color shows).

From the perspective of a part-limiting cosmology, where it is assumed that parts exist independently before entering into relations, we are not epistemically challenged in this fundamental way. We simply believe that all we need to do is to map each part, identify its essential contribution, and put it all together (typically referred to as Descartes' method of analysis and synthesis). So, if relational realism is correct, why does this method still "work"?

Typically, we search for the essential contribution through experiments. But what happens in an experiment? The idea is that we are screening off the contributions of other partners. But is this really possible? Are there no other contributors in an experiment, or are partners contributing just differently? From a relational perspective, we should say that we are not able to isolate partners; we are rather adding new ones and removing old ones. Hence, the manifestation you observe is a *different* mutual manifestation, not a manifestation of a single disposition. Hence, what the intrinsicalist ultimately does is to *infer* that, since this is different, this is the "intrinsic" contribution from partner Y. By doing so, they maintain the part-limiting idea that the bearer has this and this disposition independently of everything else. For example, when you 'scissor out' a gene from a genome and observe a different manifestation, you *infer* that this gene contributes to z. The manifestation is not evidence of a single disposition manifesting. If you are an essentialist, or just habitually defend the idea of intrinsicness, you add the metaphysical claim that this gene will perform this function *independently* of other partners (usually called conditions, auxiliary hypotheses or context).[12] This operation (including a set of unacknowledged inferential operations) allows you to assume that manipulating gene Y will consistently yield a desired (or undesired) manifestation. Through this process, science sorts out conditional claims (if p, then q) and determines whether they represent constant conjunctions (correlations) or causal relations, ultimately building theories about how nature works. This forms the foundation of modern technology in particular.[13]

The way I have described our situation may seem overly detailed and complicated, but it serves as a foundation that makes it easier to clarify what we are really doing when we manipulate and transgress limits in nature. Presenting an alternative view is crucial when we seek to change what seems to limit us through various arrangements (technology), and it highlights how these operations are often misunderstood, despite their apparent effectiveness. A central thesis of this book is that much of the dysfunction in the relationship between humans and natural limits arises within science and technology, long before we individually and politically confront 'limit issues.'

3.2.6 Summary

The previous sections provide extensive detail to clarify the ontological foundation of a relational account of limits. This level of detail may be excessive for some readers and insufficient for others. For those not satisfied, this discussion is treated in more detail in Lie (2016), in particular in Chapters 3, 4 and 5. I find this detail necessary for two reasons. First, this book builds on an understanding of limits that does not rely on an already established agreement about 'how nature works'; we need to construct an alternative that is sufficiently viable. Second, although there is a contemporary wave of relational accounts emerging, it is not enough to merely point to the general rise of 'relationalism' as an alternative to part-limiting views. We must understand the role of 'relations' in greater depth to develop a viable and operational account of limits in nature.

The natural world consists of historically created real beings (bearers) that do things. What they do depends on their manifestation *architecture (base) and* dispositional partners. This architecture counters the claim that 'the relata disappear into the relation,' but it does not determine the bearer's dispositions. What bearers do depends on other partners, and the interactions between these partners encompass more than what can be determined by the manifestation base (architecture): The relational field of dispositions holds more information than what is "extractable" from the base. The relational partners can be seen as limitations, but these limitations also make the specific dispositions possible (cf. Anaximander and Schelling).

In the context of 'relational realism,' the term 'relational' means that when something does something, it does something together with other partners. The doing already entails a relation to other partners. Without them, there is no doing. For instance, water tends to quench our thirst, indicating a stable relation between two bearers (humans and water) in this context. This relation does not arise (out of nowhere) *when* water quenches our thirst. In one sense, this suggests the existence of a *more or less* fixed 'invisible' network of relations that is ready to activate when we engage in the world. This enables me to enter into *particular* relationships with other parts of my environment (e.g., I can drink this water and expect it to quench my thirst).

When dispositions jointly manifest objects, events or processes, this enables further (iterative) dispositions at a higher level. In the case of entities perceived as things, the partners have "ossified" in a way that allows the manifestation to persist for an extended period, as seen when gases coalesce to form solid objects like planets, which (when manifested) *acquire* a new interface with a new set of partners. Generally, lower levels tend to be more ossified and rigid. There is less "liberty" the further down you go, and manipulating lower levels requires more effort to change how things manifest (see Section 3.3.1). This explains the differences we observe in the sciences. We may be more certain about 'the doing' at lower levels (physics), but this certainty stems from our shared natural history, not from a divine watchmaker or just-so *laws*.

The *possession of* dispositions depends on other partners, even if they never manifest. Consequently, we cannot establish a finite or essential description of what something is. This does not imply that x can do anything, *as x is limited by its manifesting architecture and the range of dispositional* partners. The range of *manifesting* partners is often restricted by spatial and temporal proximity. In the case of organisms, if we accept homeostatic regulations as real phenomena, there would be additional limits to consider (see Section 3.3.2).

Humans are manifesting partners. We are *in media res* when trying to understand what to do and how the world works (see also Chapter 4). We can never step outside the relational web we are part of, which complicates our scientific understanding of the world. Dispositions are not only relationally constituted but also reciprocally constituted. The task of science is not to understand how the different parts work independently of us, *but to correctly align the right dispositions with the right reciprocal partners when something is manifested.*

'Right attribution' is the key here. Our lack of understanding of this foundational precondition of scientific inquiry—and the challenges it presents—is a major reason we fail to appreciate when, why and how we transgress nature's limits, along with the consequences that ensue.

3.3 What nature can do: possible-N and possible-I

Relational realism, introduced in the previous sections, represents a third alternative to whole- and part-limiting ontologies. Limits are neither to be sought in the whole nor in the interface between the discrete parts, but in the relational bonds between the parts. At this point, however, we have not said much about the nature or being of these bonds. They might be seen as completely abstract and contingent. Rearranging them might come at no cost. The question is whether there is a certain order of the existing relational arrangements. If such an order exists, it will have a specific impact on 'what nature can do.'

A meaningful concept of limits, however, *must be one that can be transgressed while still being a limit*. If a limit disappears when transgressed, it is not a (real) limit. The relational order we are seeking must be one where limits are not only those that are impossible to transgress, but also those which are possible to transgress and where the transgression 'comes with a cost.'

In the previous sections, I argued that the doing of a certain natural entity, event or process is limited by its manifestation base, its historically evolved partners (and homeostatic regulations in the case of organisms). Here, I will introduce a fourth type of limit that encompasses the others. I call this the limits of possible-N.

We can start by asking about the true range of dispositions of a given bearer. In a particular situation, some of the bearer's dispositions are manifested, some are not. Additionally, some are normally manifested and some are not. This depends on the other partners in place. The bearer has a wider range of dispositions that for instance a scientist may explore in a laboratory. To do this, the scientist must introduce and tweak certain partners and subtract others. The banal but crucial question is: Why does the scientist even need to do this? The straightforward answer is that there is a specific way in which these dispositions manifest 'under normal circumstances.' What we observe under the contrived conditions in the laboratory is the mutual manifestation of a new set of partners that *deviates* from 'what nature spontaneously offers' (Mill, 2002, book 3, Chapter VII, §3).

An essentialist would often interpret the laboratory results as revealing the real and essential nature of the investigated partner, suggesting, 'Now we see what it does when all the other disturbing conditions are removed' (see also Cartwright, 1989). However, as we have seen, a relational realist would interpret this as a manifestation of the new set of partners (e.g., how an entity behaves in a vacuum). Thus, the lab result should be viewed as just *another* mutual manifestation. ('Yes, the potato can *also* grow faster if artificial fertilizer is added.') The laboratory does not misinterpret or construct a truth about the entity in question; it simply unveils other dispositions and the particularity of those dispositions. But are all these truths about the entity equally valid? In one sense, yes. In another, no.

The laboratory results do not capture the truth about what nature (naturally) does. What nature spontaneously does is what we try to bracket in the experiment. In order to discover what the entity does beyond its 'normal behavior,' we need contrived conditions (most systematically derived through experiments). These contrived conditions isolate and *contain* the alternative manifesting partners from those that the entity naturally teams up with. (Let's say the 'normal temperature,' 'the normal air pressure,' 'the normal genome,' 'the normal light conditions,' the normal gravitational pull, etc.) What the relational realist would say is that we do this *to investigate the broader and further* dispositions that the partner possesses, not to uncover its 'real independent nature.' This wider and alternative set of dispositions is what I term possible-I, while what nature (naturally) does (spontaneously offers) I refer to as possible-N.[14]

These two concepts are the most important concepts in this book.

To be clear, the claim thus far is that there is a realm where the set of possibilities is more restricted (possible-N) than the total range of the dispositions that a bearer might manifest *if* x, y, z, etc. How can this claim be justified? Why is it the case that we have not told the whole story when we have described the full set of dispositions that the bearer might have? Why is the purported scientific description—attempting to unveil *all* the properties of X—a misrepresenting strategy?

To understand this, we must return to the historical nature of dispositions. Now, we should note that this has nothing to do with the question of whether this history has a purpose or meaning (cf. the 'strong anthropic principle'). This history may be utterly "meaningless," but still be a specific and *particular* one in which certain dispositions tend to manifest while others do not. What we encounter in our *particular place* in this natural history are what I call 'historically evolved dispositional facts' (HEDFs) (Lie, 2016, Chapter 5.2). One such fact is—to take a random example—that we cannot break down lignin in our gut. If we could, we would be able to 'eat trees.' Evolution did in fact not take that path, even though it probably could have at some point (far back in time). But, as a matter of fact, *it did not* (while it did so for fungi). This is why we do not 'eat trees' today. Things 'might have been different,' but as a matter of historically evolved fact, they are not. This means that the way things are is not random, but the fact that they are not random is not given by any essence or eternal intrinsic set of properties,[15] rather it is given by relational natures iterated and ossified through the course of history, including and excluding dispositions that are or are not possible. The dispositional potencies that exist in nature at any point in time are relationally tied up. Things change and can be changed, but if you want to intentionally change something, there is *already* a web of relations, a bagage, that needs to be changed as well.

It is here that the significance of the relationality of dispositions begins to come into focus. What we refer to as natural or normal *manifestations* is determined by the relational web of partners that has co-developed with the entity in question. In biology, there are basic assumptions in evolutionary ecology. The chemical components in plants interact in specific ways, some of which involve the sun as a partner. This extremely complex relationship is not a random fact about plants that could have been otherwise, *given* our specific historical development of the universe. Humans *can* create contrived arrangements to replace the sun (a greenhouse), but it cannot be spontaneously replaced by another partner, and it must be *replaced correctly. This relationship which forces us to replace the partner correctly* is built into the disposition (photosynthesis) of the plant itself.

There are two things to be said about this example. Firstly, the fact that we need to replace the sun in the right way seems obvious. It goes without saying, we might say. What is obvious needs to be expressed here, however, to get to the point where we can make a case for the concept of limits that is not so obvious in the first place. Secondly, the reader may notice that I deliberately ascribe 'emit light' to the sun. This is a level-specific concept for the bearer we are talking about. It is tempting to use a more reductionistic and technical language (photons, light waves, etc.), but that would not be accurate for the level-specificity of the bearers we are talking about here.

What this means is that there are certain relational entanglements that we cannot treat as 'just another set' if we wish to manipulate the world. We cannot simply combine anything with anything else, as if the world were a collection of randomly arranged Lego bricks.[16] If every entity had random intrinsic qualities, we could suitably combine them one by one without any additional effort. We would not need to invest extra work in shielding and adding partners; they could be combined and would function indefinitely. (Something that, by the way, we would actually find hard to even imagine.) In the same way as we would not say that a car is a contingent composition of parts, we cannot generally say that about the way nature is organized either. Neither of the arrangements is accidental. The difference is that a car is largely made by humans and does not maintain itself.

I have said that possible-I represents the wider range of arrangements that reveal alternative contributions from the bearer(s) in question. A possible-I disposition is a disposition that is not possible to manifest for free. A disposition that requires human effort and maintenance to manifest. We may observe that bearer X can do actions f, g, h and i, but through certain interventions, we learn that X can also perform actions j, k, l and, to our surprise, can also do r if we (humans) add partners Y and Z and remove partners A and B. The problem with our understanding of limits today is that we often fail to recognize this difference. The final word about the potency of X is not set by any suggestive possible-I range. The actual range is also determined by possible-N. Not only in a conceptual sense, but also in a practical way. To make a plant grow faster (in the laboratory), we need to shield off, add and maintain new partners, and this rearrangement of partners must be done in certain ways and with certain efforts *because of the already established range of possible-N*. The range of possible-I is not only wider, but also determined by the narrower range of possible-N. This might seem obvious, but if it is, why is there hardly any recognition of a possible-N order (Lie, 2016, Chapter 1.2.1 and 2)?

At this point, we should clarify that the dispositions themselves are real, regardless of whether they are possible-N or possible-I. Possible-I dispositions are no less real or somehow inferior to possible-N dispositions when asserting truths about reality. In one sense, this means that humans can "create" many truths. Consequently, possible-I truths are *not* socially constructed; they do not misrepresent reality, as Nancy Cartwright (1983) would argue. Bearers possess a wide range of dispositions, and those manifested ("created") in the laboratory are neither more true (essentialism) nor less true (constructivism). What would be untrue is the claim that idealized conditions unveil what is *really* true about X.

What I suggest here is that truth is not a particularly interesting issue (Lie, 2016, 161), and the success of science is not a miracle (cf. Putnam, 1985). What is interesting is naturalness—how do natural systems truly work? To create our possible-I systems, we must understand how the possible-N system operates. We cannot simply isolate and utilize possible-I manifestations without considering how the possible-N realm works. I will clarify this interdependence further in the next section.

3.3.1 What are limits—the order of nature

In this section, we will draw some conclusions about what limits in nature are, based on the premises developed throughout Chapter 3 thus far.

We may already see why the possible-N realm does not refer to an order that denies the possibility of change and manipulation. We are neither discussing an impeccable whole-limiting order nor an immutable order of external laws (necessity). On the other hand, the possible-N realm does represent a type of order. The internal relations within this realm are not random. For example, geese partly navigate by the stars; most life forms on Earth, as they exist today, depend on specific temperature gradients; fluorine lacks one electron in its outer shell, making it highly electronegative and reactive; and hens are originally adapted to forest habitats in South and Southeast Asia, where they used their claws to dig in the soil for food. The act of digging in the soil is not an intrinsic property of the hen; it is a relational disposition. The limits imposed by this order are neither absolute nor contingent. Where there are dispositions, there are also webs of relations that may not be visible or necessary, but are nevertheless present. This relational order enables our existence, yet what makes us possible also limits us. For instance, geese cannot navigate (by the stars) when the sky is cloudy, as they depend on a clear sky to see the stars. When I attempt to exceed these limits, it incurs an expense.

According to the Endurance Running Hypothesis, when early humans (*Homo* erectus) developed the ability to sweat, lost most of their body hair and evolved longer legs to outrun (and overheat) larger prey, these adaptations made a certain life-form possible—but they also limited the range of environments in which this hominin could thrive. As our ancestors evolved, they became more adept at transgressing these limits by creating clothing and simple housing. However, this also incurred an expense—the creation and maintenance of clothing and housing. Clothing and housing do not simply arise on their own. The simple fact that such *costs* exist strongly suggests that the possible-N order consists of 'real phenomena' rather than being purely socially constructed.

If such a difference exists, it should also be reflected in our understanding of truth. Something can be true about what is possible-N and something can be true about what is possible-I. The truths regarding possible-N limits are not the same as those regarding possible-I limits. If we fail to make this distinction, we risk asserting that only those limits that cannot be changed (physically, practically or economically) are the only limits that exist in nature. Consequently, we might arrive at conclusions such as: 'We thought it was impossible to create a synthetic organism; we now know this is possible' or 'No one believed it was possible to create a fusion reactor; we now know this is possible.' So, what did we think before changing our minds? Many things, but many of those thoughts can be categorized as 'products of our minds,' or social constructions. The fact that these limits could actually be changed and transgressed seemingly suggests that what we previously believed to be real is not real after all. From this, we wrongfully conclude that our prior beliefs must have been social constructions, akin to: 'Before, we didn't believe women could compete in ski jumping; now we know they can.' Failing to distinguish between possible-I and possible-N puts us in a situation where we tend to regard immutable limits as the real limits of nature, while deeming the mutable ones as (eventually) socially constructed.

The claim in this book is that there is a difference between transcending social constructions and transcending the possible-N order. The indication that we have surpassed the possible-N order is that we need to establish new partners that require maintenance for the preferred manifestation to occur.

Thus, the argument here should be straightforward and not particularly controversial. It is a rather banal claim: who would deny that clothes do not make themselves? The idea that humans must 'work in their own sweat' was already stated in the Book of Genesis (Genesis 3:19). However, if this banal truth is accepted, then science or any semi-authoritative source of knowledge should also acknowledge that there is a 'natural order in nature' (imposing expenses on the trespasser), which would be a rather controversial claim. Numerous scholars take pride in debunking the 'myth of naturalness,' one of the first being John Stuart Mill (1923). Nevertheless, if we maintain that there is an order—neither strictly necessary nor purely contingent (viz. modal dualism)—that has emerged through the evolution of our universe, shaping and constraining us, then the implications of the seemingly banal notion of expense reveal themselves to be quite profound.

To clarify what we have established so far: The suggestion is that we are transcending the limits of nature when *what we do requires a setup of new dispositional partners for the desired dispositions to manifest*. We recognize these as new and not possible-N when we need to *maintain* the set of these partners. This means it is indeed possible to surpass limits in nature. Some limits may be difficult to transgress because they require an enormous setup of contravening partners (fusion reactor), while others may be easily surpassed (like making clothing).

Just as naturalness is not normative, the limits in nature are also not normative. There is nothing inherently wrong with crossing limits. The main point of this book is to explore and understand what it means to go beyond the limits of nature when we understand them in this way. I argue that in many cases we should transgress the limits of nature as little as possible, but not because limit transgressions are (morally) wrong. The reason lies in the implications of these transgressions and whether the rationale behind them is sound. My point is that what we aim to achieve by exceeding limits is often not realized. In fact, considering what was said about frontier culture, I believe we often attain the opposite of what we seek (see Chapter 5).

3.3.2 Plasticity

The relational web of possible-N adaptations is flexible. Organisms are homeostatic beings capable (within certain limits) of regulating critical functions independently of external influences to maintain their integrity.[17] They also exhibit developmental or phenotypic plasticity, allowing for various life-history strategies, morphs and changes in cellular functions in response to alterations in the external environment. On a population level, there are numerous abilities to change rates of offspring production, generation time, etc. These types of plasticity are temporally reversible. Irreversible temporal plasticity occurs when populations undergo changes in their gene pools and adapt to a shifting environment through mutations, genetic drift, epigenetic inheritance, niche construction and horizontal gene transfer. Ecosystems (or biological communities) also demonstrate a certain type of "plasticity," as they can revert along a successional track toward some kind of "endpoint" when disturbed.[18]

To assess the plasticity of what I have termed possible-N, ecologists often use the concepts of resistance and resilience, which can also apply to other aspects of nature.[19] Resistance refers to the ability to tolerate disturbance, while resilience is the capacity to recover from it (Holling, 1973).[20] For example, a particular population of species X might be highly vulnerable to a specific disturbance yet still recover quickly. An ecosystem, such as the Barents Sea, may be easily disturbed by intense and unsustainable management but can return to a similar state when left undisturbed. Similarly, the human brain may be affected by a concussion but can recover fairly quickly.

Plasticity is what we refer to when discussing limits *as* tipping points. Tipping points denote critical thresholds where a system shifts into a different state or trajectory, often irreversibly. At a tipping point, a system can no longer resist or recover from disturbances that alter its dynamics. Examples include when the body can no longer regulate blood sugar levels (e.g., diabetes), when a population loses herd immunity, when sea temperatures reach a level that causes coral organisms to die, 'or when a certain concentration of greenhouse gases triggering feedback loops leading to further emissions—such as methane released from thawing permafrost.'

While one may debate whether these concepts apply to specific units (e.g., ecosystems, organisms or populations)—that is not the primary concern here. The key point is that we (individuals, cultures, institutions, etc.) have some notion of 'plasticity in nature.' For instance, I can drink alcohol even though it is poisonous, we can chop wood in the forest without risking the collapse of the forest itself, and governmental bodies can regulate the decimation of certain fish populations 'within safe limits.' However, the problem is that there is no clear philosophical agreement on what plasticity means.

An illustrative example is the invasion of sea urchins in the coastal areas of Northern Norway since the 1970s, initially believed to be driven by overfishing (Norderhaug et al., 2020). The urchins consume kelp—specifically, the stem of the kelp—leading to the creation of 'urchin deserts.' The habitat that once protected the fish preying on the urchins has also been severely diminished, as kelp provides shelter for their fry. This complicates the recovery of the original ecosystem. Who would have imagined that this outcome would result from overfishing along Norway's shorelines decades ago?

In an effort to restore 'the original system,' there have been attempts to drastically reduce the urchin population in larger areas. These efforts can be interpreted in various ways:

1 Humans caused the system to collapse, and we must restore it to its original state.
2 Urchin deserts have significantly depleted coastal fish stocks, which the local human population (and others) regrets, so we need to restore them to revive the relationships that supported their economic prosperity and cultural practices.
3 Regardless of the cause, the changes that this system has undergone are neither inherently good nor bad; it is simply another ecosystem. If we wish to restore it, it would be because we *choose* one system over another.

In the first interpretation, we have clearly transgressed the plasticity of the system; it should be in a certain state, and that is not the case now. In the third interpretation, there is no such fixed state. We can choose which state we want, and the idea of plasticity, therefore, becomes a misleading concept. In the second interpretation, there is a specific human goal involved. The plasticity of the system has been transgressed relative to this goal. The original assumption about plasticity (in the 1960s)—that "we can decimate these populations as much as we do now"—was flawed, and now we see the unintended consequences of that assumption.

The misleading notion about plasticity is found in the third interpretation (which I will attribute to ecomodernism). We cannot simply choose which system we want if there is already an established system. If we demand that this view be consistent and *meaningful*, it would imply that we might as well could have a system where we decimated the fish population, as was done in the 1960s, *without risking the subsequent sea urchin invasion*. But, the urchin desert is not *just another system*; it is a *result* of what was done to the historically evolved interconnectedness that was originally there. The state of the new system (urchin dessert) is not contingently related to the previous system. In other words, the (degree of) plasticity is not contingent upon what we want, but on possible-N.

The third interpretation is correct in suggesting that we could have had it both ways, but that would require a huge subsequent effort to control the invasion. The possible-N dispositions do not disappear; they are merely concealed. All efforts to control the system (in having it both ways) would thus represent an alternative manifestation of the real and transgressed plasticity. So, if there is no such thing as full plasticity (3rd interpretation), we should ask whether the fishermen at that time would have *chosen* to decimate the population and face all the efforts to stop the invasion. In the following chapters, I hope that a further exploration of this understanding will clarify this point.

3.3.3 Limits and burdens

As argued in Section 3.3.1, the transgression of limits in nature comes with and is defined by burdens. This is not the type of burden that follows from Adam and Eve being thrown out of paradise, but burdens related to *containers*: their production and maintenance. There are two main burdens related to transgressions. The first burden relates to the work required to establish and maintain the possible-I containments. The second pertains to environmental problems related to these contained arrangements.

The first type of burden originates from the relational nature of nature. If nature were a set of distinct and independent tiles acting only in accordance with their intrinsic nature, we could rearrange the tiles in ways that serve our intentions without burdens. As previously argued however, the "tiles" are already interconnected in specific ways, and if we want to deviate from this historically evolved order, we need to establish a new set of relationships and maintain them. This is typically what we refer to as technology or technological arrangements. We don't consider the regulation of sugar levels in the body or the geological drivers of ocean salination as technology. The effort required for this is commonly referred to as 'work' (Karl Marx), which serves as the justification for what we recognize as 'private property rights' (John Locke).[21] Work refers to thermodynamics and encompasses all sorts of energy sources necessary to create and maintain a particular order. In Section 3.1, we acknowledged that living beings need to *act* in certain ways to exist and persist, countering the tendency of nature toward higher entropy. If humans want to go beyond the biological work required to counteract this entropy, we must provide mechanisms that oppose nature's tendency toward possible-N. For example, a farmer growing potatoes needs to weed the land and add fertilizer to ensure the potato plants manifest as desired. In such cases, we *protect and contain the possible-I from the* possible-N. If not protected, the possible-I arrangement will eventually revert to a certain dynamic possible-N state. A typical example of such a state would be a successional end state occurring in an ecosystem.[22] A famous example is the Chernobyl area, post-meltdown.

The fact that 'possible-I containers come with burdens' is not far from being a truism. The fact that we find it completely reasonable to pay for a product or that stealing from someone is wrong reflects this fact.

The second burden concerns environmental issues. By accepting that possible-I requires protection from possible-N to sustain its functional existence, we accept – at least in practice – that there is such a thing as a possible-N order (or similar). It is therefore arguably reasonable to assume that possible-N also needs protection from possible-I. In such cases, we assume a symmetry between possible-I and possible-N. This symmetry, however, *is rarely recognized* or valued, and there are several reasons for this.

The first and most obvious reason is that we do not have to accept a symmetric relationship. To make the possible-I work, we *need to* contain it *from* the possible-N. Otherwise, it will not function as intended. We need to protect the combustion chamber from overheating and the metal in a car from corrosion, for instance. However, the reverse is not true; we do not *strictly* need to contain the possible-N from the possible-I for the possible-I to work (Lie, 2016, Chapter 5.4.1). This is why I call this an *asymmetric relationship*. A certain industrial facility serves as a *de facto* acknowledgment of the existence of a possible-N order; it's a container. Yet it remains possible to deny that possible-N needs protection from possible-I. This is particularly advantageous for companies aiming to minimize costs, as it enables them to reduce containment despite broader effects on possible-N. When this denial is enacted, we encounter what economists refer to as externalities. The environment is forgiving; the effects are shared by and dissipated into the commons, and the causal pathways are often complex enough to obscure responsibility.

A second reason arises from our general assumption that the existing order is contingent—that it could just as easily be different. This perspective encompasses a broad range of views, from scientists who hold a modal dualist view of nature to the truck driver who spills oil onto the ground *without* hesitation. We could refer to this phenomenon as the asymmetry of contingency. In general, I argue that the precautionary principle faces a fundamental tension with the assumptions of dominant scientific ontologies.

The principle asserts that when there are non-trivial possibilities of harm, the burden of proof falls on the producer to demonstrate safety. However, if we adopt a passivist, part-limiting ontology—one that views nature as a contingent arrangement of discrete parts—then the possibility of unforeseen adverse effects is effectively 50/50 in all instances. In this framework, any intervention (e.g., substituting sugar with aspartame) could be perceived as having an equal probability of causing harm or ensuring safety. If this is the case, the burden of proof becomes epistemically unresolvable: How can one decisively prove safety in a world where any change implies an inherent 50/50 risk? Consequently, the precautionary principle risks becoming either paralyzing or selectively applied, depending on how this ontological assumption is interpreted. When nothing stands out as more or less likely from the outset, the onus is on society A liberal socitety in particular to prove non-trivial effects *in every single case*. One could argue that this is not the case in practice, but what would be the *ontological* justification for that? If there is no possible-N order and everything is subject to free recombination, what should guide our assumptions?

By taking possible-N as our point of departure however, we can assess *the* likelihood of harm upfront based on how far and in which direction the possible-I arrangement deviates from possible-N. If there is an ontological justification for possible-N, our hypothesis would not assume that there is a 50/50 chance that aspartame is safe as these molecular structures have not been part of our biological evolution.

Thirdly, we should consider reasons where the asymmetry is acknowledged but not fully understood or cared for. A typical example would be *unintended* consequences. As early as 1986, Ulrich Beck (1992) described in *Risk Society* how the production of risks has come to overshadow the production of goods and prosperity, leading to fundamental changes in the priorities of society and politics (see Section 5.6). Others have expressed concern about how science operates as a co-supporter of ignorance through its practices (e.g., Wynne, 1992).

The fundamental problem of ignorance is that the range of what x can do is wide and it is hard to predict 'which *reciprocal* partners will do what' when a new container is put into use. When we are ignorant in this sense, we don't even know what we should have known, i.e., unknown unknowns.

Non-relational ontologies, such as reductionism, have a particular perspective on unintended consequences that arise from ignorance (Lie, 2016, 172). If a certain behavior can be determined solely from the constituent parts—such as nano-silver—we may not need to worry about the 'relational effects' at higher levels, as they are presumed not to exist. From a reductionist perspective, we believe we already know the consequences in advance: 'This herbicide only kills these specific weeds,' or 'The edited gene construct only alters the production of protein F.' Consequently, the precautionary principle may not even apply, since we can assert that x only does f.[23] In this way, scientific reductionism might inadvertently support the development of technologies that fail to recognize possible interactions and effects on broader systems. We may refer to this *modus operandi* as essentialist asymmetries.

Fourthly, the asymmetry may be accepted to some degree when we acknowledge trade-offs. Typically, one might accept certain impacts on the possible-N web if the consequences are deemed acceptable. In this context, 'acceptable' means that the benefits outweigh the known costs. There are various issues associated with cost-benefit analysis that I will not address here. It is important to recognize that economists require quantifiable data and that the relational nature of possible-N rarely appears in these numbers.

Lastly, the asymmetry is accepted for a good reason: our deviations are within the range of the plasticity of possible-N. The challenge is that it is not entirely clear what this plasticity entails in each specific situation. It depends on the scale in time and space and the actions of people elsewhere. The atmosphere can absorb CO_2 emissions from, let's say, all the inhabitants of the USA, but if everyone emits at the same per capita rate as American citizens globally, we are likely to face irreversible effects on possible-N. A useful rule of thumb is that rapid, large-scale changes typically outpace the adaptive capacity of most living organisms. Rapid and extensive changes will favor possible-N entities that have shorter generation times.

3.3.4 Burdens—not human-nature dichotomies!

Up until this point, I have argued that possible-N and possible-I are opposites, which would seem to reinforce the nature/culture dichotomy. However, if we think this through, we will realize that this framework essentially argues against human/nature dichotomies. Any possible-I arrangement would consist of a combination of possible-N and possible-I partners. When the emphasis shifts toward possible-I partners, there are relatively fewer possible-N partners directly involved, and vice versa; however, there is never an arrangement devoid of possible-N partners. There is no such thing as an artificial X. A general principle in this gradient is that the greater the number of possible-I partners needed to achieve a desired manifestation, the more burdensome it becomes in terms of work. The excessive energy demand in our modern societies—where transgressions of possible-N are prevalent—illustrates this point. This is precisely why we should be concerned about the burdens imposed by environmental problems. For this very reason, we should view environmental issues as the reflection of deviations from possible-N.

Possible-N partners represent the "labor" we receive for free. To the extent that we, as biological organisms, and our possible-I systems *depend* on this free labor, we should be concerned whenever these "free dispositions" are compromised. It is therefore not the case that any arrangement of nature's components is as good as any other. Certain arrangements imply more burdens. Extensive deviations from possible-N arrangements require more work, and they also demand additional effort later on, wherever the asymmetry has been enacted and environmental problems created. The possible-N partners already involved in our existing systems must be replaced by possible-I partners, ultimately increasing the workload. This is how containers swell—from neat, controlled structures into bloated, insatiable organs that are always hungry and always growing.

The truth is that I am not presenting anything particularly new or controversial here. It is well understood that when a farmer needs to replace "free rainfall" with costly irrigation systems due to climate change, it also results in more work. What is novel is the framework, the emphasis on the relational essence of our being, and the insistence on the existence of a natural order. I contend that we cannot reach this conclusion without a notion of possible-N. While the conclusion itself might not be very controversial, the premises are, which is why the seemingly convivial perspective we arrive at requires a book to carefully outline the entire argument.

3.3.5 The size of the container

If possible-I activity results from intentional activities, there are two ways in which the transgression of limits in nature does not create environmental problems.

1 The activity is within the given range of plasticity of the possible-N system.
2 The container is symmetrical.

In the previous section, I asserted that any possible-I arrangement requires a new configuration of dispositional partners, where the possible-N partners are more or less included. As a starting point, we can understand this as a combination of possible-N and possible-I partners, but the idea of 'composition' needs further attention. Is it a combination of distinct possible-N partners and distinct possible-I partners? In what sense are the possible-N partners substituted and removed? I argue that the possible-N partners are still present, but they are "represented" differently within technological systems. For instance, the steel that supports a bridge manifests, so to speak, the possible-N dispositions it counteracts. This insight has significant implications for the broader argument in this book: we should not claim that our possible-I arrangements have *overcome* the limits of nature. Those limits persist; they are simply converted into possible-I partners. In fact, it is precisely the constraints imposed by possible-N dispositions that compel us to engineer our systems in the *specific* ways they are constructed. A combustion engine requires a cooling system because of the thermal and chemical properties of iron and other materials. In colder climates, we must add antifreeze to water because water has a physical limit to how cold it can get before freezing. These examples show that nature's limits are not abolished by technology; they are manifested differently. In this sense, it is still true that technology shapes nature, *but it is also true that nature shapes technology*. The cooling system does not eliminate the thermal limit; it 'expresses it.' What I will refer to as 'limit arrangements' are precisely these configurations in which possible-N constraints are iterated through technological form. However, it is not the case that such arrangements are natural in the sense of possible-N. Possible-N is the reference point from which such arrangements can deviate more or less.

If we accept this view, it is also easier to see that the question of the size of the container pertains to the *optimality* of this relationship. There are two reasons for this.

First, if our containers are intimately dependent on possible-N partners—not just in the sense of different identifiable ecoservices but also in the sense of being 'expressed in a different way', it is also evident that we depend on them. To assume that asymmetric containers are viable is therefore self-defeating in a profound sense. If, by building a bridge out of steel, we also altered the possible-N dispositions represented by the steel in a way that gravity was suddenly doubled, the bridge would collapse. While this is an admittedly far-fetched example, it illustrates the kind of dependence we are trying to escape when deploying asymmetric containers.

It is only by accepting this premise that we can move to the next step in our argument for optimality. If possible-I and possible-N depend on each other in the intimate sense described above, we also need to assess the genuine *efficiency* of our interventions differently.

In a possible-I arrangement, we want to obtain something—a manifestation. We want to produce electricity, we want to produce salmon, we want to grow potatoes, etc. There are different ways to achieve the same manifestation, and in evaluating what we should do, we can combine possible-N dispositions and possible-I dispositions in various ways. I would argue that the most efficient way to operate is the one that harnesses as many possible-N dispositions as possible—because they contribute for free. It is the wind in the air, the river's fall from the mountains, and the self-renewing aquatic system with its surrounding currents that refresh the water in which the salmon farm is embedded—these are the sources of surplus. In contrast, possible-I partners contribute only as much as they cost, which aligns with the insight of the second law of thermodynamics. While this might seem obvious, there is a catch: If we do not care about the symmetry, we may harness these dispositions *as if they did not belong to a possible-N order*.

An example could be salmon farming in Norway, which emerged in the mid-1970s. It started with containers that were just able to protect and contain the manifestation of the intended possible-I dispositions from possible-N: a facility with a net placed in what the farmer saw as a convenient location (close to home, a place where no one protested, a place close to roads, etc.). In the beginning, many of these farmers went bankrupt repeatedly. There was rarely an infrastructure for distributing the products, and the genetic profile of the salmon was not well adapted to the open, yet contained, habitat of the farmed stock. Additionally, there were numerous diseases and health issues. As these problems were solved along the way, *the container* grew. The ways in which the possible-N system reacted to the initial possible-I container *forced the growth of possible-I partners* for the fish farmer. 'Nature comes in package deals.' Despite these obvious feedback patterns, we may safely assume that the focus and intentional activity of the fish farmer *continued to be directed toward keeping the container as small as possible*. Because it is also obvious that the more partners needed, the more expensive and less profitable the operation becomes. As the business grew, with expenses following, this led to a more capital-intensive operation requiring the consolidation of commercial operators that were better able to benefit from scaling. The solution was 'bigger is better,' not to acknowledge that we need to find an optimal balance between possible-I and possible-N. Thus, an aggressive expansion in terms of locations and facility size followed, resulting in a second wave of problems. The number of issues associated with this business today is tremendous.[24]

I use this example to illustrate how the container grows as the business grows and how every step toward a more wide-reaching container, also containing effects on possible-N, is routinely rejected but eventually (indirectly) accepted. In all these steps, the main concern is not possible-N, but the possible-I container. The intentional powers are always leaning toward minute concerns for the possible-N system. It is very rare to see a business that evolves from a perspective where the possible-N system is the primary concern from the start (see also Chapter 4). There is little or no acknowledgment that we should seek to develop optimal limit arrangements.

The main point is that the business starts with a plan that eventually turns into a defensive strategy to protect against what appears to be isolated obstacles—'limits that need to be overcome.' In a part-limiting view, there is nothing given but contingent parts freely open to rearrangement—i.e., parts (limits) you believe can be dealt with as they arise. From the viewpoint of a relation-limiting perspective, you would explicitly anticipate the need to place the container within a certain already existing order for it to work. Seen from the relation-limiting perspective, it may also occur that this order can be used systematically to our advantage. One might consider an optimal size and density of the stock or whether kelp farming or mussel farming can be combined with salmon farming (to reduce eutrophication). All such adaptations would yield less profit, but also fewer possible-I partners. The only way that the use of more possible-I partners becomes more profitable is if the fish farmer pays the same "asymmetric" price for the possible-I partners as the consumer pays for the salmon. No one is paying for the externalities when they pay for the product. These are only paid indirectly through environmental problems, the effects of which on people are disconnected from those who gain the benefit of a lower price. Thus, the size of the container is not determined by what is optimal, but by an economic system that favors those who can maximally benefit from the asymmetry. (You cannot escape protecting possible-I, but you can escape protecting possible-N.) Thus, unless there are rules and laws that counteract this tendency, there will be less technological development that aims at optimality between the possible-I and possible-N partners. Those who seek optimality will not survive in the capitalist economy.

3.3.6 Where to go from here

Limits in nature present an exceedingly complex issue. For those who seek clear-cut boundaries to define limits on human expansion and activity, the reality is that no such limit exists. Limits pertain to what x can do. A limit of 450 ppm might be the right number in terms of how much greenhouse gases we can let into the atmosphere (before we, e.g., enter into a hot-house scenario), but it is only the right number given a certain fixed set of manifesting partners. The number might be 500 ppm, if we at the same time emit x number of aerosols (making aerosols a resource – a product that can be sold on the market). Limits are relational. The idea of using limits as a knock-out argument against the economic-growth optimists is therefore counterproductive, as such claims will repeatedly be proven incorrect, at least in the short run (see also Sections 1.5 and 6.6).

The relational nature of limits may actually reinforce the argument that there are no inherent limits in nature. Throughout this chapter, I have contended that this perspective is valid if we assume that nature is endlessly rearrangeable. This view can be supported by passivist part-limiting ontologies that assert the world consists of separate components, contingently arranged from the outset and open to ongoing contingent reconfiguration.

I have argued that it actually works to treat nature as if it were an assembly of contingently arranged parts. Nature indeed offers a much broader range of possibilities than we typically recognize. However, this 'liberty' comes at a cost; it is only attainable if we accept a corresponding level of work and/or environmental challenges. But why should this be a concern? It is not an issue as long as we forget the fact that our initial goal in these rearrangements was to enhance our freedom. Limits in nature and freedom are interconnected, but not in the sense that we automatically gain more freedom by transgressing nature's limits. Our relational bonds can be transformed but never overcome. Our dependencies can only be replaced with other dependencies. What I can do depends on how other intraconnective parts behave when I act. There is no such thing as a free lunch. If we do not accept this as a foundational premise in our quest for freedom, we may end up becoming enslaved by our efforts. This topic will be explored further in Chapter 5.

It is clear that determining who decides on limit arrangements is challenging. For example, when we consider the size of a container, we soon realize that these containers are interconnected. A car cannot encapsulate all its effects on possible-N. Whether the car is symmetric depends on its production methods, the construction of the road system, its energy source and the processes involved in manufacturing and delivering that energy. Decisions regarding these matters are not made at an individual level, which will be discussed in Chapters 5 and 6.

At the same time, as individuals, we encounter limits in nature and make decisions about whether to transgress them. This will be the focus of the next chapter.

Notes

1 This is not an accurate statement since the second law applies to closed systems (and where, e.g., the universe as a whole would be such a closed system). Life on Earth takes place in an open system where the sun provides 'free energy.'

2 This is a debated question that I discuss at length in Lie (2016). See especially chapter 6.4. The idea of succession is far from the same as that Clements developed in 1916, but it is still a concept that is accepted and taught in biology courses. See also Jax, 2010, 75.
3 Some readers may find it useful to compare the account given here with Bruno Latour's actor-network theory. As this book is not about ontology as such, I will not do that myself.
4 E.g., Mumford, 2004, 170. In Sections 2.3 and 2.4, I described how Aquinas's concept of law evolved from referring to the consistent behavior of 'an Aristotelian substance' to describing the deterministic behavior of all material things. This transformation ultimately contributed to the modern idea that laws are external to the things themselves, imposing order upon them from without.
5 However, this does not mean that the modern-day dispositionalist subscribes to a teleological worldview. The potency is not a potency to realize the form of a substance.
6 Some may recognize the familiarity between 'more than it is' with Graham Harman's notion that 'objects have a surplus' or are 'withdrawn' (Harman, 2002).
7 A little ahead of myself, I will condense the reasons as follows: The first reason is the 'always packing, never traveling' objection (Molnar, 2003, §11.2). A second reason is that neutral monism is less vulnerable to the general objection raised against relationalism: that the *relata* (part) disappear into the relation. By asserting that what has been called the categorical base by categoricalists and dualists is really the manifestation of dispositions at a lower level, we also overcome the traditional disagreement between process and object ontologies. Manifestations can be durable (objects), momentary (events) or "cascading" (processes). See Lie, 2016, chapter 4.6. The last and most important reason is that we don't have to argue that dispositions are intrinsic to the bearer, (pace Mumford's relationalism), which would imply that dispositions are held *independently of everything* else. I argue that if we accept that dispositions are intrinsic (held independently of everything else), we must deny that they are relational. We cannot have it both ways. This is where my view (relational realism) diverges from Stephen Mumford's relational view of dispositions (or what he prefers to call powers). I distinguish between "belonging to" (adherence) and "intrinsic to" (see Lie, 2016, chapter 4.5.3). In Lie, 2016, chapter 4.1, I argue that the concept of intrinsicness causes relational theories to revert to non-relational and quasi-relational theories, leaving them open to reductionist strategies.
8 If you are a reductionist, you will not accept drinkability as a real disposition of water, as 'drinkability is not entailed in the molecular structure of water.' This goes back to intrinsicalism, deeming drinkability as an extrinsic ("relational") property. Hence the term relational realism.
9 To this we may also add what is sometimes called the relational turn found in other sciences (see, e.g., West et al., 2020).
10 Phrases like 'is responsible for' and 'regulates your mood' are other common misconceived and disguised ways to describe what serotonin *does*.
11 Those familiar with debates in the philosophy of mind might recognize this as a counterargument to Jaegwon Kim's causal closure argument (Kim, 1992).
12 Essentialism and habitual intrinsicalism often do not differ significantly; you don't need to explicitly adhere to essentialism to conclude that Y does Z independently of other factors.
13 I am not suggesting that this is unrecognized in the philosophy of science. The same issue appears, for example, in the distinction between necessary and sufficient conditions and in the problem of underdetermination, although the ontological context differs.

14 I admit that these are not very sexy terms. I use these terms to keep a distance from what is usually called natural and artificial, and which have a heavy load of connotations. I also use them to explicitly emphasize what is important here, namely potency. It is sometimes convenient to say that: Well, that is possible, but not possible-N, for instance. I also want to underline that these expressions do not hold a normative content. If something is possible-N, it does not entail that it is also good. (Lie, 2016, chapter 5.7). The observant reader may also have noticed that I have subtly introduced a concept of naturalness here. This concept is highly controversial and needs justification, which I will provide in the next section (see also Lie, 2016).

15 This assertion taps into a larger debate between essentialists and Humean contingency theory (see Mackie, 2006).

16 In Lie (2016), I refer to a conversation between Eric Drexler and Rick Smalley where Smalley tries to explain to Drexler why his Lego-model of atomic activity is completely flawed (pp. 71–72). The conversation can be found in Baum (2003).

17 For some readers, the assumption that organisms are homeostatic beings may seem premature or even presumptuous, as this remains a contested issue. One way to mitigate this controversy is to note that physiology and medicine standardly use the terms 'hypo' and 'hyper'—terms that only make sense if we assume some kind of balance point relative to which these deviations are defined. Our claim doesn't need to be stronger than this. Personally, however, I believe that homeostasis is a principle central to biology. Organisms can react to external influences and thus maintain their integrity through feedback processes and diverse reactions based on the information these processes convey. The organism as a whole influences its parts to maintain optimal conditions such as temperature, salinity, blood sugar, pH, calcium and oxygen levels. I recommend J. Scott Turner's book, 'Purpose and Desire: What Makes Something "Alive" and Why Modern Darwinism Has Failed to Explain It' (Turner, 2017), for those interested in exploring whether there are stronger claims to be made about organisms and homeostasis. Particularly illuminating is his analysis of how the concept of vitalism has been mobilized to create conceptual confusion, thus reinforcing the prevailing (passivist) paradigm in physiology. (Few wish to be associated with vitalism as it is defined, but theories of homeostasis are not necessarily vitalist.)

18 Again, some may argue that I am making presumptuous claims. The idea of an end-state has been debated since Frederic Clements introduced the concept of climax community in 1916, based on Henry Cowles' ideas of succession. Despite this disagreement, it is safe to say that ecologists generally agree that *succession* is a real phenomenon (see Lie, 2016, chapter 6.1.2).

19 A reference point that is sometimes used is the 'relative stable period of Holocene' (Rockstrom et al., 2009). This demonstrates the need for a reference point when discussing these matters, and I have argued in the previous chapters and 2016 for why I think that possible-N is the most robust one. For a general criticism of applying any reference point, see Jørgensen, 2014; Kareiva & Marvier, 2011; Marris, 2011, and below.

20 Holling uses a wording that is slightly different.

21 Marx's notion of 'arbeit' is typically translated as 'labor,' but since I want this term to also encompass all forms of energy transfer needed to produce and maintain a specific order, more closely aligned with the use of 'work' in thermodynamics, I prefer the term 'work.'

22 As noted previously in footnote 14, ecologists do not agree on how to describe this end state. For any such state, it is also essential to keep the general principles of resilience and disturbance from Section 3.3.2 in mind. There may or may not be a return to a *similar* (though not identical) possible-N state, depending on the degree of disturbance and the resilience of the system. In my opinion, this disagreement stems from a debate in ecology that gained significant traction following the introduction

of a new paradigm, which is referred to as *The Ecology of Disturbance and Patch Dynamics* (Pickett & White, 1985). This dispute can be traced back to the disagreement between ecosystem ecologists and population ecologists. Daniel Botkin (1990) referred to this as "New Ecology," while others refer to this paradigm as 'chaos ecology' (Lie, 2016, chapter 6.1.2).

23 While this *idea* follows from an essentialist theory based on reductionism, it does not apply in practice. For example, there are protocol restrictions on the application for most products that seek to enter a commercial market, such as required field experiments. However, in certain scientific fields, such as gene technology, we are led to believe that new gene-editing techniques, primarily CRISPR-Cas9, have made the technology extremely precise and safe. The underlying idea is that "side effects" arise from inaccurate methods of manipulating specific genes within the genome. But what do we mean by inaccuracy? Is it possible to determine the function of individual genes at all? From a relational realist perspective, I would say 'no,' unless we can guarantee that other components in the genome remain unchanged. This would be nearly impossible, given that "single genes" potentially interact with all parts of the genome, especially if we consider some top-down influence on genomic activity. Furthermore, it is difficult to predict how the precise alteration of coding activity will affect higher levels, such as how a protein will fold and how this folding may influence relational behavior within the cell. The point I want to convey is that as long as science fosters the belief that we can safely manipulate nature based on essentialist frameworks, it will continue to contribute to unintended effects down the line.

24 Risikorapport norsk fiskeoppdrett 2024 | Havforskningsinstituttet (hi.no) (English summary).

References

Armstrong, D. (2005). Four disputes about properties. *Synthese*, *144*, 309–320.

Baum, R. (2003). Nanotechnology: Drexler and Smalley makes the case for and against 'molecular assemblers'. *Chemical and Engineering News*, *81*, 37–42.

Beck, U. (1992). *Risk society: Towards a new modernity*. Sage Publications.

Bird, A. (2007). *Nature's metaphysics: Laws and properties*. Oxford University Press.

Bird, A. (2016). How the powers ontology has overreached itself. *Philosophical Studies*, 13.

Bird A. (2016). Overpowering: How the powers ontology has overreached itself. *Mind*, *125*, 341–383.

Botkin, D. (1990). *Discordant harmonies: A new ecology for the 21st century*. Oxford University Press.

Cartwright, N. (1983). *How the laws of physics lie*. Oxford University Press.

Cartwright, N. (1989). *Nature's capacities and their measurement*. Oxford University Press.

Cartwright, N. (1994). *Nature's capacities and their measurement*. Oxford University Press.

Ellis, B. (2001). *Scientific essentialism*. Cambridge University Press.

Ellis, B. (2002). *The philosophy of nature*. Acumen.

Harman, G. (2002). *Tool-being: Heidegger and the metaphysics of objects*. Open Court.

Harré, R., & Madden, E. H. (1975). *Causal powers: A theory of natural necessity*. Blackwell.

Heil, J. (2004). Properties and powers. In D. Zimmerman (Ed.), *Oxford studies in metaphysics* (Vol. 1, pp. 223–254). Oxford University Press.

Holling, C. S. (1973). Resilience and stability of ecological systems. *Annual Review of Ecology and Systematics*, *4*(1), 1–23. https://doi.org/10.1146/annurev.es.04.110173.000245

Jax, K. (2010). *Ecosystem functioning*. Cambridge University Press.

Jørgensen, D. (2014). Rethinking rewilding. *Geoforum*, *53*, 282–288. https://doi.org/10.1016/j.geoforum.2014.11.016

Kareiva, P., & Marvier, M. (2011). *Conservation science: Balancing the needs of people and nature*. Macmillan.

Kim, J. (1992). Multiple realization and the metaphysics of reduction. *Philosophy and Phenomenological Research*, *52*(1). https://doi.org/10.2307/2107741

Ladyman, J., & Ross, D. (2007). *Everything must go*. Oxford University Press.

Lewis, D. (1986). *On the plurality of worlds*. Blackwell Publishing.

Lewis, D. (1997). Finkish dispositions. *The Philosophical Quarterly*, *47(*187).

Lie, S. A. N. (2009). *Naturalness reconsidered: Dispositions, ethics and ontology*. University of Tromsø Press (Doctoral Thesis).

Lie, S. A. N. (2016). *The philosophy of nature: Rethinking naturalness*. Routledge.

Mackie, P (2006). *How things could have been: individuals. Kinds and essential properties*. Oxford University Press. (Or Clarendon Press. Oxford?)

Marmodoro, A. (Ed.). (2010). *The metaphysics of powers: Their grounding and their manifestations*. Routledge.

Marris, E. (2011). *Rambunctious garden: Saving nature in a post-wild world*. Bloomsbury.

Martin, C. (2007). *Mind in nature*. Oxford University Press.

Martin, C. B. (1994). Dispositions and conditionals. *The Philosophical Quarterly*, *44*, 1–8.

Martin, C. B. (2008). *The mind in nature*. Oxford University Press.

Mellor, D. H. (1974). In defense of dispositions. *Philosophical Review*, *83*, 157–181.

Mill, J. S. (1923). *Nature: The utility of religion and theism*. Longmans Green. (Original work published 1863).

Mill, J. S. (2002). *A system of logic: Ratiocinative and inductive*. University Press of the Pacific. (Original work published 1843).

Molnar, G. (2003). *Powers: A study in metaphysics*. Ed. S. Mumford. Oxford University Press.

Mumford, S. (1998). *Dispositions*. Oxford University Press.

Mumford, S. (2004). *Laws in nature*. Routledge.

Mumford, S. (2007). Powers, dispositions, and properties: A causal realist manifesto. In R. Groff (Ed.), *Revitalizing causality: Realism about causality in philosophy and social science*. Routledge.

Norderhaug, K. M., Nedreaas, K., Huserbråten, M., & Moland, E. (2020). Depletion of coastal predatory fish sub-stocks coincided with the largest sea urchin grazing event observed in the NE Atlantic. *Ambio*. https://doi.org/10.1007/s13280-020-01362-4

Pickett, S. T. A., & White, P. S. (1985). *The ecology of natural disturbance and patch dynamics*. Academic Press.

Putnam, H. (1985). *Philosophical papers, Vol. 2: Mind, language and reality*. Cambridge University Press. (Original work published 1975).

Quine, R. (1974). *Roots of reference*. La Salle III, Open court.

Rockstrøm, J. et al. (2009). A Safe Operating Space for Humanity. *Nature*, *461(*7263), 472–475.

Schrödinger, E. (1944). *What is life? The physical aspect of the living cell*. Cambridge University Press.

Turner, J. S. (2017). *Purpose and desire: What makes something "alive" and why modern Darwinism has failed to explain it*. Harper One.

West, S. et al. (2020). A relational turn for sustainability science? Relational thinking, leverage points and transformations. *Ecosystems and People*, *6*, 304–325. https://doi.org/10.1080/26395916.2020.1814417

Wynne, B. (1992). Risk and social learning. In S. Krinsky & D. Golding (Eds.), *Social theories of risk* (pp. 275–297). Praeger.

4 Coming across limits in our everyday life

4.1 Introduction

I have titled this chapter "Coming Across Limits in Our Everyday Life." However, the title may be somewhat misleading regarding the analysis presented here. What I am about to argue is that we are typically *not* stumbling upon or coming across limits as if they were simply 'lying around,' ready to be perceived and considered. Rather, they are hidden, leading us to the question: how do these limits even occur to us at all?

To overcome limitations in nature, these limits must somehow surface. They need to present themselves as some kind of issue. For most living creatures, however, limits simply exist and do not register as limits. Some may even question whether the term 'limits' accurately conveys the right meaning if they are not experienced as such. For instance, I do not typically perceive gravity *as* a limit. Often, they may be perceived as "conditions" rather, if perceived at all. If this is the case, how do we come across limits at all, and how is technological development possible? Technological development is the process where you *velop* out of something you are *veloped* into: but if you don't know or notice your "invelopments," how do you de-velop?

To deepen our understanding, we need a proper starting point to examine these types of experiences. For reasons that I hope will become clear, there is no better starting point than Martin Heidegger's phenomenological analysis of our 'being-in-the-world.'

4.2 Introducing Heidegger

In a part-limiting view, we perceive other parts as obstacles to our goals. The idea is that humans pursue objectives, and in doing so, we encounter certain parts that we consciously assess, deciding whether to accept, avoid or overcome them. However, Heidegger describes this situation quite differently. Most of the time, we are 'absorbed in the world.' As goal-directed beings, we are (and must be) situated within a relational whole when interacting with other parts. This 'being-in-the-world' is what enables our encounters with individual parts in the first place.

DOI: 10.4324/9781003307372-5

Martin Heidegger was a student of Edmund Husserl, and his philosophy is regarded as a significant contribution to the phenomenological school initiated by Husserl. In 1927, he published his most famous work, *Sein und Zeit* (*Being and Time*), which is the focus of our discussion here.

Husserl was skeptical of the empiricist explanation of how we experience the world around us. His account serves as a radically different contender to British empiricism, primarily regarding how we genuinely experience something. Husserl argued that the world does not "really" appear as sensory data that is re-presented, mapped, summarized and made significant in the mind, as Hume suggested. Instead, the mind is nothing over and above its intentional *content*, which is always *about* something meaningful. For instance, if it concerns a cup, it is *not* primarily about (the sensory data of) color, shape, handle or weight, sequentially organized by the mind as a cup. While I can use my cognitive capacities to analyze the cup in terms of color, shape, etc., that is not how I *experience* the cup initially. So, how am I able to recognize it as a cup? Husserl's idea was that we can experience the "essence" of things, and his aim was to reconstruct this process so as to bid the world an alternative to the empiricist view of experience (cf. the logical empiricists who insisted on grounding scientific theories in 'data' during the same period).

Heidegger agreed with the aim of phenomenology but disagreed with several basic assumptions that Husserl took for granted. One of these assumptions is that we are positioned as cognizing subjects who may operate on our own experiences (eidetic reduction and phenomenological reduction) to enhance our understanding of the world. Heidegger contended that our experiences of the world are so intertwined with it that an analysis of this kind would simply 'destroy the phenomena.' For the most part, we exist among and in the midst of a world of "objects." Heidegger sought phenomenological descriptions that arise 'from within.' How does the world appear when you are a part of what you are experiencing (and describing)?

Heidegger believed that any given description cannot escape being a part of something not included in the description. For example, the scientific description of let's say a cup will be part of the scientific world, along with its concepts, ontologies, theories and instruments, none of which are explicitly mentioned in the description itself. However, this does not represent the cup's world, so to speak. How would a description look from the perspective of the world to which the cup belongs? This is Heidegger's question. Starting from this point, Heidegger doubts that we can see the cup as a cup *because* we grasp its "essence." Rather, we gain insight into what the cup is by participating in the same whole as the cup. This is what Heidegger refers to as "seeing it from itself." We recognize the cup as a cup because there is a table, a mug, a chair, a plate, a fork, a person… *there* as well. In other words, it is all these other things of which the cup is a part that 'make it a cup' and make me experience it as a cup, rather than my ability to grasp its 'essence.' In this sense, there is no cup 'in itself.' We can see the resemblance between Heidegger's understanding of the human-made world and our previous relational account (Chapter 3) of the world beyond humans.

So, why is this a useful starting point for understanding how limits 'show up' in our lives? The main reason is that any examination that begins with the premise that we are constantly confronting and taking a stance on 'the limits of the world around us' would be fundamentally flawed. Practically speaking, there exists an almost endless array of limits that I do not notice and to which I do not relate, whether voluntarily or involuntarily. A description that assumes we are already embedded in these limits, in a practical and nontheoretical manner, would therefore be preferable.

While there may be issues related to Heidegger and his philosophy, many agree that his exploration of our practical engagement with the world is a significant contribution to the philosophical tradition in general and perhaps the least controversial aspect of his philosophy when considered in isolation.

4.3 Being-in-the-world

According to Heidegger, humans are always *caring* (*sorge*) about something (1992, 238). We care because we *care about our being*. Not first and foremost our sheer biological existence, but about *who I am*. The being of man (*dasein*) is not fixed. We are not objects, but "objects" that constantly relate to ourselves, our own being. I am a man, but more importantly, I am a man *in this and this particular way*. I may be a medical doctor, but my being is not about being 'a doctor,' but *the way* I am a doctor. I constantly choose *who I am and have to be* (which is also why Heidegger is often characterized as an existentialist). Every day I get up in the morning to care for and 'maintain myself.'

Being goal-directed agents in this way, 'things around us' stand out as meaningful or significant (except in situations where *angst* occurs). However, the way to describe this Heideggerian goal-directedness is not by describing how we consciously and rationally pick a goal and *choose* the means to achieve it. Rather, as we have seen, Heidegger thinks that we are 'always already' in the world where the things (tools) *already* carry a sort of goal-directedness. Instead of the term 'human,' Heidegger uses the term *dasein*, which is supposed to designate the circumstance that we are not an isolated thing-subject among other things, but a being that is 'out there,' already related in a caring mode of being.[1]

The world that *dasein* encounters is a world of tools (*zeuge*). The philosophical tradition has had a prevalence for thinking that the world consists of things (substances); trees and birds, chairs, atoms, genes, cells, etc.[2] Tools are Heidegger's equivalent to *things*, and he claims that tools are more primordial than 'things' (1992, 96). The forest is wood, the sun is light and warmth, the stone is a hammer and so on. *Dasein* is *in* this world, not in the way water is in the glass, but as a familiar absorbed dweller (Heidegger, 1992, 79). *Dasein* is (for the most part) familiar with this world exactly because it is a functional world that makes *dasein*'s care possible.

Tools are always in-order-to (*um-zu*) something. *They point beyond themselves*, to other tools. The lawn mower points to the lawn, to sharpening tools, the electric grid and so on. The totality of tools points 'toward-which' something, which is the work (*werk*), the main aim of the operation (Heidegger, 1992, 99). Thus, the *meaning or significance* of the lawn mower is not to be found in the lawn mower itself, but in its relation to the lawn and other tools, and finally, the work itself. Every "thing" point beyond themselves and have no intrinsic being. Human intentionality does not sit inside the human mind, therefore, but is rather laid out in a world of "things" that already signify ("intends"). *Dasein* finds itself *thrown* into this relational whole of signifying tools (*zeug-ganze*), which makes up a meaningful background for everything in this world. This means that to understand a certain *dasein* and her actions, you need to understand the relational world of which she is a part in order to see and understand 'what makes sense' for this being. *Dasein's* world is not a world of facts but is something that can only be understood through interpretation, because *dasein* is continuously interpreting herself.

This is a rough outline of the reasons that Heidegger would likely put forward to counter the idea of the 'cup in itself.' Today, we would typically say that the cup 'must be understood in its context,' which has become a common expression, partly thanks to Heidegger. The point being that the transparency of the world does not arise from the detached theorizing mind, but from being-within this structure of relations 'already there' (Heidegger, 1992, 79). Without this world of significant relations, nothing would make any *sense*, which is to say that nothing makes any sense in itself, independently of us. This is also how Heidegger challenges the description of the subject seen as an agent that freely, consciously and rationally chooses its goals and the means to achieve them. Human agency always unfolds within a relational whole in which goals and means are already embedded and where these relations form the precondition for any meaningful action.

4.4 If-then: how can something be different?

Heidegger's account of how we relate to the world around us seems to poorly align with the observation of our frontier culture, where transgressing limits seems to be 'a mode of being.' First, humans exist within a nexus where goals and means are already more or less given. Secondly, humans do not encounter nature; they only encounter tools that already possess given meanings. Third, transcending this world of significations would imply entering a realm of meaninglessness.

However, there is a way for us to explain how we are transcending the structural whole that Heidegger refers to as the world, also on Heidegger's account. To understand his thinking on this topic, we need to examine *how* we relate within this world.

Heidegger argues that there are essentially two ways to relate to tools. The more primordial of these is what he calls ready-to-hand (*zu-handen*). This occurs when we use tools in a familiar manner (1992, 98). However, tools can sometimes shift to being present-at-hand (*fûr-handen*). When tools are present-at-hand, we perceive them as independent objects ('things'). This mode of relating partly allows us to step outside our embeddedness and study the beings in question independently of ourselves and our use.

As previously stated, we might typically expect that a theoretical attitude toward the subject matter—such as mapping all the tools and their relations or studying them carefully one by one—would yield the best understanding of what these things truly are. According to Heidegger, however, the best way to understand things, i.e., the nexus of tools (a particular world), is by *using* the tools. This understanding arises *only* when tools are ready-to-hand, where we use them without attending to the tool as such. We *subordinate ourselves* to the tools. Heidegger refers to this type of 'inattentive attention' as *circumspection* (*umsicht*). Different tools emerge (are brought closer) in accordance with their significance specified by the tasks that revolve around the continuation of our being. When I use the keyboard on this computer, my attention is not on the keyboard itself, as it was when I first began typing on a typewriter (when it was present-at-hand). It is by *not* attending to the tools that the world opens up. The more I subordinate myself to their assignments and allow the tools to 'do their job,' the better I understand what the philosophical tradition has referred to as a thing. An expert is someone who can make the tools entirely disappear in her handling by *following* them (cf. Dreyfus & Dreyfus, 1986). This occurs when the reference-nexus (the nexus of the in-order-tos) of the tools functions as a kind of horizon or context for the dynamic use of specific tools, and when there is enough skill to handle them in 'flowing' consent with their "bending" (*bewandnis*) (Heidegger, 1992, 115, footnote 2).

From this description, we can identify at least two aspects of *dasein's* experience. Firstly, Heidegger concurs with G. W. F. Hegel that things (or rather concepts in Hegel's case) perceived as independent (i.e., in themselves) are abstract entities. It is the whole, the context, the nexus of tools that render the thing *concrete* and define what it is. Secondly, to access or understand the beings in question, we paradoxically need to engage with and use the 'things' in a manner that makes them fade from our attention. It is when the "things" become tacit or withdrawn that the relevant background understanding develops, *and it is this type of ('noncognitive') background understanding that can disclose the being of the beings in question*. In an 'ideal' situation, one might imagine that this intimacy would be so profound that there would be no real difference between the mind and the world outside the mind.[3] Again, the world is a web of relations that does not reveal itself (as Heidegger would say) to the external observer meticulously noting their appearances.

The reason this particular analysis garnered attention initially relates to the epistemological debate on 'access to the world,' which we have touched upon earlier. Heidegger's account drew attention because it challenges the notion of privileged access to the world that science has traditionally held. He questions the conventional subject–object relationship as the foundation for endorsing a particular scientific method. However, this is not our focus here, and that is why we will not delve into, or reflect on, the differences between tool-totality, the world and the relationship between what Heidegger calls the world and reality/nature at this point.[4] What we want to explore further is how something akin to a subject–object relation is possible in the first place. How does the presence-at-hand occur?

Heidegger (1992, 103) asserts that this type of relation arises:

1 When tools are destroyed or rendered unusable,
2 When something obstructs their use (obtrusiveness) and
3 When a tool or tools are missing.

He refers to this as a *break* (*ein bruch*). The tools lose (more or less) their 'in-order-to character' and present themselves as 'things.' The entire nexus of tools, the assignment-totality, becomes *conspicuous*. It becomes conspicuous precisely *because* our everyday dealings with the world typically occur in a ready-to-hand mode. Our conspicuous (and sometimes incredulous) gaze allows us to suddenly 'see' how everything is interwoven (*meldet sich die welt*). A gestalt shift occurs. For example, when driving a car that suddenly breaks down on a cold winter day, the car becomes conspicuous; it is merely a cold object standing there, revealing to our conscious mind all the tasks that the car *does* within the larger nexus of assignments.

At this point, it is essential to recognize that sheer presence-at-hand is merely an intermediate state that *cannot endure*. If it persists, one might call it paralysis, and if it continues for an extended period, it would even be a sign of anxiety (angst). When objects become present-at-hand, *dasein* focuses on solutions involving other tools (e.g., a towing truck) and returns to the ready-to-hand mode. This is the situation where our explicit *if-then* reasoning can replace our circumspective *in-order-to* "reasoning." In other words, it is normally only when the 'world breaks down' that we can start to think about alternative ways to connect the parts and transcend what may occur as limitations.

This conveys a crucial insight: a technology is functional/good/useful *if it can be integrated into ready-to-hand circumspection*. In this respect, there is not any crucial difference between a wooden spade and a spade made from steel. If the user is only familiar with or knows about wooden spades, it can be made ready-to-hand through a learning process. Our relationship with technology is primarily determined by *whether it is ready-to-hand* or not—not by its "objective" qualities or efficiency. Every tool has obtrusive and nonaligning aspects, no matter how efficient, shiny or smooth it may be 'in itself.' As long as these limitations can be masked by the ready-to-hand circumspective use, the "real" obtrusiveness does not matter (see Section 4.7). If you cannot make it work in a ready-to-hand mode in a way that allows it to disappear from your attention, *it will not be a proper tool*. It will merely be an object that prevents you from attending to the task (*werk*) at hand. This indicates that it is not the tool that facilitates alignment, but rather the ready-to-hand mode of comportment. I refer to this aspect of technology use as the *neutral alignment of technology*.

4.5 When something is missing

Dasein's being is a being in time: past, present and future. *Dasein* is thrown into the world and makes future projections (*Entwurf*) about its possible being. These projections, concerning *dasein's* worldly existence, are shaped by the nexus of assignments *opened up* by *dasein's* use of tools. *Dasein* compares itself with its possibilities based on how the world appears within the totality of assignments. For something to be real, it must be possible (Heidegger, 1992, 184–185). One might assert, while lying on one's back and dreaming about space travel, that it is possible to travel to the moon. While traveling to the moon is indeed possible, it exists only in what Heidegger calls a factual (*tatsächliches*) sense: everything that is physically possible is considered possible (cf. possible-I).

When viewing humans as free and independent within a broadly liberal framework, a familiar challenge arises: distinguishing between 'wishful possibilities' and 'real possibilities.' This distinction poses no problem for Heidegger. Real possibilities are those that occur within the world of the assigning nexus. To the extent that humans are free, choices are made based on an understanding of *what is happening* (i.e., ready-to-hand circumspective comportment) in relation to what 'you are and have to be.' This concept is what Heidegger refers to as facticity (*Faktizität*). *Dasein's* primary way of being 'in the possible' is, first and foremost, in the mode of *expecting*. *Dasein* expects something within a range (*Spielraum*) of possibilities, which is partly constituted by the nexus of tools. If you *are* a passenger on a ship, the possibility of entering the wheelhouse to take over the steering does not occur to you. In other words, as a passenger (holding the identity of a passenger), the possibility of taking over the helm is not a 'real possibility,' even though it could logically and physically be possible. Some things are expected, while others are not.

Heidegger's unveiling of the nature of real possibilities is more meticulous than I can fully address here. But why is this important? For humans to exceed or transcend certain limits in nature, there must be a situation where an individual or a group recognizes an alternative way to achieve a goal. Analyzing how possibilities play a role in our lives is therefore crucial. When a tool is used, it represents the only possibility and enables *dasein* to be determinate in its comportment. This is why it would be problematic if *dasein* continuously expected other possibilities in its dealings. More importantly, Heidegger's understanding of possibilities offers a clear view of how the possibilities and limits of nature are mediated by tools. For instance, if you are a biker, possibilities and limits reveal themselves through 'the bike.'

I have emphasized that the subordinate use of tools (ready-to-hand) is a significant aspect of our relationship with technology. If a tool becomes present-at-hand, it is, by definition, nonfunctional, regardless of how efficient or "perfect" it is. A tool has a neutral alignment, meaning that if it is or becomes present-at-hand, it is also 'unusable' or, rather, *im-possible* to use, irrespective of the cause. From this, we can infer that there exists a contradiction between how we use tools and how transgressing limits must occur (from a first-person/phenomenological perspective). Again, to 'go beyond limits,' which requires *dasein* to envision and expect other possibilities, seems impossible. Nevertheless, there appears to be no reason to doubt that humans are expanding their range of possibilities by exceeding limits.

In this context, we could: (1) abandon Heidegger's analysis and assert that it is simply incorrect; (2) completely shift our perspective to, for example, an analysis of the capitalist system and its role in our expansion into the natural world; or (3) adhere to the Heideggerian account and modify it in a way that seems to resolve the contradiction.

Everything would, of course, be in vain if we did not opt for the latter option, and we will proceed by expanding on what it means when something is 'missing in the workshop.'

Consider the workshop as a metaphor for the technological system that enables *dasein* to address tasks that matter. What is missing could either be something that (1) was there before, or (2) something that never existed but *could have been there*. If the missing tool was previously present, it is straightforward that there will be a 'moment of presence-at-hand' where circumspection halts, and *dasein* must either stop working or find alternative solutions (e.g., locate the missing tool or find a replacement).

In the case of option 2, the problem is more complex. When *dasein* identifies a possibility that is not already present, it could be due to having used a tool on another occasion (borrowed from a neighbor), having seen the current item elsewhere (e.g., in an advertisement), or simply envisioning this 'other possibility' in their mind (as in the case of a potential invention or daydreaming). In all these instances, *the existing* tool(s) become present-at-hand, depending on *dasein's* ability to vividly bring the missing tool into focus. The *obtrusiveness* of the current tool, which was *always there* (but hidden by its ready-to-hand use), becomes conspicuous, making it difficult for *dasein* to subordinate to the task at hand and use the tool in a ready-to-hand mode (e.g., 'I always thought shifting gears manually was a hassle anyway'). The limits in nature, mediated by technology but obscured by ready-to-hand comportment, become constraints. *The obtrusive properties of the wooden spade become apparent when dasein brings the steel spade close, leading to an understanding of its limitations.*

This brings us to the issue of 'bringing close' (*entfernen*). As mentioned, this could relate to psychological profiles. Human traits such as patience and sensitivity (to resistance) can make a difference. Additionally, we must consider the degree to which there are offers and the ability to access them. People living in a capitalist market economy encounter a categorically different level of *offers* than those in households relying primarily on local resources (just as a rough example). Likewise, the *ability* to access or obtain the tool in question influences one's capacity to remain in a ready-to-hand mode. An economically affluent person is likely to consider other possibilities more frequently than someone for whom most relevant options are merely a dream. (Although, at times, wishful thinking may also influence efforts in ready-to-hand use, depending on psychological profile.)

If we agree that technology is *neutrally aligned*, it becomes evident that three factors—psychological profile, offers and ability—are crucial for analyzing technological progress and the transgression of limits in nature. When the *other possibility* illuminates the current possibility and highlights the *obtrusiveness* of the existing technological item, it transforms into a 'thing' that stands out as unusable. *Dasein* can then either 'force itself' back into circumspection or address (now or later) the new task of acquiring the item that initially created the presence-at-hand situation. In the last instance, I experience *a need* for X, for instance, a robot lawn mower. In the first instance, you 'make yourself content' with your current lawn mower. I shall refer to the first case of realignment as *reconciliation* and the second as *expansion*.

This description offers an alternative to the more common narrative of viewing humans as inherently Promethean: that we are fundamentally transgressing limits by our very nature. The belief that this tendency is inherent to us relies on the assumption that we *actively* seek transgression. However, the analysis presented here posits that we primarily strive for immanence, a ready-to-hand relationship with the world. Humans do not actively seek expansion; rather, we appear to pursue *unity and oneness* with our surroundings. If this is the goal or 'nature' of *dasein's* whereabouts, transgressing limits is merely one of several ways to address *problems in the ready-to-hand relationship*. Transgression is not part of our nature (if there ever was or is such a nature). However, transgressing limits (expansion) is, admittedly, a tempting strategy that is hard to resist.

4.6 Technological development and the assessment of efficiency

According to Immanuel Kant, our rationality can serve as an instrument for selecting the best means to reach a goal (hypothetical imperatives) and can also make us capable of governing our pursuit of goals in accordance with universal principles we give to ourselves. The degree to which a means can achieve an end is measured in efficiency. For instance, one may recognize that riding a bike is more efficient than walking the distance from point A to point B.

There are two critical strands discussing the topic of efficiency.

A vivid critique, rooted in philosophical romanticism, centers around the claim that rationality has collapsed into one at the expense of human ends and autonomy. As has been noted, Heidegger (1977), Horkheimer and Adorno (1972), Jacques Ellul (1964), Herbert Marcuse (1964) and to some extent Jürgen Habermas (1984) have argued that developments in modern Western societies have largely come to be defined more by efficiency itself than by human ends. While efficiency typically involves finding the most effective means to achieve a goal, they assert that it has become an end in itself. When asked, 'Why should we do X?' the answer often is simply, 'Because it is more efficient.' Where there is broadly no invitation to further questioning and the use of our autonomous will, one might conclude that efficiency has become a goal in itself. Thus, rationality has become instrumental (rationality).[5]

The other critique centers around the question of whether what is portrayed as efficient truly is efficient. For example, is industrial agriculture genuinely more efficient than small-scale farming when considering external inputs, water use, soil compaction, erosion, CO_2 emissions and more? (See, e.g., Weis, 2010). This raises specific assessments related to the question of "the size of the container" (Section 3.3.5).

So, how are these questions about efficiency related to the question of limits in nature? As many would recognize (also from what has been said so far), there is a close relationship between limits and efficiency. Limits in nature define the boundaries for achieving certain goals in specific ways. For example, there is a limit to how fast a plane can travel from point A to point B due to various constraints. By altering one or more of these conditions—such as fuel properties, air resistance or the plane's weight—one can achieve the goal *more efficiently*. With technology X, I am able to reach farther and accomplish more with the same effort. I can achieve the same or new objectives by changing some of the manifesting partners. That is: using better means. This is what we call efficiency. I can water my plants with a hose rather than carrying water back and forth. I can listen to the news on the radio while preparing food instead of sitting down to read the paper. I can provide instructions to ChatGPT and make it write an essay for me instead of writing it myself. A lumberjack can clear a site of trees in one day instead of one month, using a logging machine instead of a horse. A pill can cure a medical condition that I otherwise would have to endure.

Now, if we return to our Heideggerian analysis, we will see that the *other possibility* is not 'just another option'; it is a possibility that is more efficient. If it were less efficient, it would not "threaten" the current use of tools, and *dasein* would not be disturbed in its ready-to-hand engagements. A wooden spade does not trigger a present-at-hand mode of comportment if *dasein* already uses a metal spade. This is why we refer to it as technological development. When *dasein* can achieve the same goal with less effort, we say that *dasein* is more free in this particular sense.

This description helps us better understand how technological development unfolds as it does.

1 First of all, the analysis emphasizes that the general goal that brings technological development about is not expansion or transcendence, but rather immanence. *Dasein* focuses on the *other possibility* as a strategy to return to ready-to-hand comportment. This new possibility makes the current one obtrusive while it *simultaneously presents a solution to the problem*.
2 If *dasein* has brought the other possibility close and it is available, the easiest way to return to the ready-to-hand mode is by retrieving this new tool. The reason is that the current tool has become an "unusable" present-at-hand object, especially when the new one is *perceived as* more efficient. This explains the certain type of "gravity" associated with new possibilities, as the new option can render the current tool present-at-hand (when it is viewed as more efficient). My emphasis here is not on the better qualities of

the new tool, but on the 'unusability' of the old one, and therefore the hardship of reconciliation and staying with a present-at-hand situation. Over time, expansion will outcompete reconciliation. A stubborn individual who rejects using a smartphone will likely feel increasingly weighed down, as her resistance requires effort that was not necessary before the new possibility emerged. This may help explain what is often referred to as the technological imperative.[6]

3 It is not the fact that the new tool is more efficient than the current one that creates the present-at-hand situation, but rather the perception of it. *Dasein* can be misled into believing that X is more efficient compared to Y.

4 The assessment of efficiency is often not a rational overall evaluation of whether X is genuinely more efficient than Y. When *dasein* operates in her ready-to-hand mode, tools are interrelated, and their use and usefulness are connected to the entire nexus of assignments. Whether a possibility stands out as real (and not merely factual) depends on the entire assignment-nexus within the tool-totality. A nail gun is only more efficient if you have electrical power. While this may limit technological development, it also enhances it, as *dasein* does not consider efficiency as such but efficiency in terms of what I will refer to as *nexus efficiency*. When a *break* occurs and *dasein* enters a presence-at-hand mode, she predominantly evaluates new possibilities within the *narrow* context of the existing network of assignments already there.[7] Thus, technology that might not be genuinely effective (or ethically sound) *from a broader perspective* will be deemed efficacious because the evaluation occurs within this nexus. In a particular situation, buying a car "in itself" might not be the best answer (to efficiency), all things considered, but it may nevertheless be the most nexus efficient thing to do.

5 The introduction of new technologies is not always limited to one-to-one efficiency substitutions for old ones; instead, it often opens new areas of discontent and perfection. When you 'can do something about a problem,' you also "create" new problems. (I will refer to this phenomenon as contingency extensions). While plastic surgery is beneficial in many cases, it also generates a plethora of 'defined conditions' (convertible to products) such as 'asymmetric nostrils,' 'gummy smile,' 'small lips,' 'puffy eyes,' 'undefined jaw,' and so on. Similarly, the growing capacity to monitor our own biometrics (smartwatches) creates notions and conditions like 'low step count,' 'poor sleep efficiency,' 'elevated resting heart rate,' 'low oxygen saturation,' and 'inconsistent activity patterns.'

Again, what occurs is that new possibilities illuminate the existing conditions we live with, transforming them into chosen possibilities. Something you can or should do something about. What is becomes, what if. Most moderns have acknowledged that such capacity extensions are not unconditionally beneficial. What seems to open up new possibilities might turn into entrenched pressures, such as when parents of children diagnosed with attention-deficit/hyperactivity disorder (ADHD) feel irresponsible if they do not adhere to a medication program, or when new technologies that make cars safer also shift responsibility for injuries or fatalities that may occur in a crash. Our enhanced ability to address situations redirects responsibility away from "the given" and toward individual agency.

What should usually take place, according to our Heideggerian analysis, is that the new possibilities integrate into the assignment totality and become a part of our daily circumspective concern. However, surrounded by an expanding array of new possibilities, it becomes harder to subordinate to the *facticity and resoluteness* of ready-to-hand comportment. A general and more or less strong and undefined condition of present-at-hand takes over, shifting the focus from what I can do and what I possess to what I cannot do and do not have. When this is the case, life becomes a defensive list of wrongs, making moments of appreciation and gratitude a rarer circumstance.

4.7 The role of sensitivity

What has been termed the postmaterial thesis teaches us that consumption beyond basic needs occurs because the desire for "new stuff" taps into postmaterial goals such as identity, self-expression and autonomy (Inglehart, 1977). If you want to sell a product that "no one really needs," you should find a way to associate the product with these postmaterial values. This explains why we consume beyond our "real" needs, why technological progress continues and why economies grow in countries where most of the population has satisfied needs that extend far beyond basic needs.[8]

Another way to explain the growth in needs is through the transition of certain nonbasic needs (satisfying wants) becoming necessities. Computers, cars and cellphones, once considered luxury items and symbols of conspicuous consumption, have now become part of the category of necessity goods. In reference to our current analysis of the being-in-the-world of *dasein*, we can understand this transition as a shift where an item changes from being somewhat outside the nexus of tools to becoming integral to that nexus, also by altering the nexus itself. While a computer in 1982 was a tool that was, in *some* cases, more "efficient" than a typewriter, it is now an inescapable part of a nexus that defines our technological landscape. The product (tool) has created its own infrastructure and established itself as indispensable through the expansion of the potential nexus of tools of which it is now a crucial part (cf. Kondratiev waves and the Collingridge Dilemma). When economists express optimism or pessimism about the economy, they assess our current position in relation to these cyclic transitions. Are we in the midst of a wave? Are we at the end of one? If so, is there a new wave on the horizon? The very idea of *a knowledge* economy—where education, especially universities, play a crucial role—is based on the notion of technological waves (see, e.g., Lundvall, 1992).

A third avenue for understanding technological expansion involves considering the role of *sensitivity*. Sensitivity pertains to what I referred to as the psychological profile in Section 4.5. I argued that individuals differ in their ability to engage with and remain in the ready-to-hand mode. Some may become easily frustrated, while others may be of the kind that easily brings other (more efficient) possibilities(existing/or noninvented) close or fails to.[9]

While these characteristics relate to individual human beings, we can also observe that societies become 'statistically more sensitive.' In this context, the sensitivity we discuss is more directly linked to supply and ability. In a prosperous society, there are typically more options available and more ability (purchasing power). In practice, this means that alternative possibilities are presented more frequently, and the ready-to-hand circumspection is *broken* more often, leading to the statistical prevalence of expansive realignment.

As previously noted, the process that occurs when circumspection is broken, and an alternative is chosen as a solution to the presence-at-hand situation (expansion), is not primarily rational. The assertion here is that even a slight improvement in nexus efficiency could disrupt circumspection, and in that case, it is not rational considerations but the level of sensitivity that correlates with reconciliation and expansion. An improvement that you would not even consider (bring close) some years ago may trigger expansion today. If this is the case, we can conclude that the level of sensitivity has increased. More generally: As your sensitivity increases, so does your ability to notice and identify problems and areas for improvement. The crucial point is *whether* we are drawn out of our ready-to-hand mode or not. Not the purported improvement itself. Since there is always the potential for an improvement in nexus efficiency, no matter how infinitesimal, *it is impossible to establish a rational limit to technological development*. In other words, technological development elevates the level of sensitivity, while the increased sensitivity facilitates further technological advancement—indefinitely.

4.8 Limits to our exploration of everyday limits

Connecting the analysis presented here with limits in nature is not a straightforward undertaking. Expanding the efficiency and range of technological performance does not necessarily equate to transcending natural limits. While there are numerous instances where technology addresses social issues, it is debatable whether transcending limits in nature inherently involves technology, understood as tools.

Discussing limits in both nature and the social domain raises complexities. For instance, it is not clear that a tool developed to kill more efficiently in warfare or a tool designed to recognize faces via surveillance cameras—ostensibly to resolve certain social issues—are cases of 'transcending limits in nature.' On the other hand, one could make the case that such inventions do actually involve the transgression of certain natural limits. The creation of an algorithm, for example, entails manipulating electrons, which may also cross specific boundaries. Enhancing a rifle's capability to shoot at greater distances might require a new type of alloy. It is not always easy to pin down when and where we should say that our technologies represent transgressions of limits in nature. However, there are inventions that expand human capabilities beyond the technological realm, which we will not address here. Cases like the establishment of a new religion may broaden human capabilities but they fall outside our current scope.[10] Similarly, situations where individuals move to a new house or choose a different partner due to a situation where 'the current one has become conspicuous,' are interesting but not relevant to our inquiry.

Another critical issue is the distinction between technique and technology. Analyzing the development of technique may be more challenging within this framework. The improvement of a skill, such as breaking a record in the 100-meter butterfly, does not follow the mechanisms outlined here, though perhaps the development of a new *method* for growing potatoes does.

In other words, our exploration of how we relate to limits in nature, as discussed in this chapter, should only apply to cases that fall under the Heideggerian framework. This means that it concerns limits in nature mediated through our use of tools.

A more pressing issue is whether the actor-level we have adopted as our analytical starting point is the relevant 'actor' when discussing modern technology. If the framework only applies within the Heideggerian framework, one might argue it is outdated. Indeed, individual humans often appear to have little or no control over the course of our current technological development. (Winner, 1977, 27). Is it truly individuals who transgress the limits of nature, rather than organizations and collectives of which humans are a part?

To answer this question, we must distinguish between the processes of invention and consumption. In the process of invention, entities like Google, Nike or ECA Group—Naval Drone Systems operate in a ready-to-hand mode, occasionally generating new ideas that lead to new products. However, in a capitalist system, competition compels companies to actively invent more efficient tools. If you don't innovate, you die.

The question arises whether these companies invent based on the breakdown of the ready-to-hand mode. I believe this is not the case. The reason is that it is the invention itself that is the task (werk). Thus, their inventions are typically not a result of a break in the ready-to-hand mode. While inventors may of course experience moments when the ready-to-hand mode breaks, this disruption typically pertains to the tools used *in* their innovative work. On the other hand, the inventor invents tools, relying on the assumption that *consumers* (whether individuals or collectives) will experience the disruption. Their product is based on the trust that this will happen and even within a collective, someone must propose and make the decision that renders the current tool obsolete. There would for instance be identifiable people that come up with these types of decisions when, e.g., a philosophy department purchases iPads for grading students. If the *break* process does not occur, the product will typically not sell.

Therefore, I contend that the transgression of limits in nature "is local." We, *as* humans or societies, do not transgress natural limits when there is a 'breakthrough' in science. The transgressions following such breakthroughs must, as we've seen, fit the local nexus-worlds of *daseins*, and it is in those local instances that limits are crossed. If many *daseins* share the same world, the product will likely 'fit more worlds' and achieve wider sales. When transgressions of limits in nature take place, they take place locally by the use of tools that combine into a possible-I arrangement, which needs maintenance or is the unintended result of such an arrangement. Such arrangements are most often something we are a part of and not something that has grown out of the choices made by an individual. (Typically, people who seek 'off-grid lifestyles' would in part be exceptions.) This assertion, however, does not take away the significance of the process that must take place, i.e., the process of expansive realignment.

What we seek to understand is how technological development and with that, the transgression of limits in nature, is possible at all.[11] This question is particularly relevant in a time when many people feel that technological development is taking a direction that may not benefit either humanity or the natural environment. Beyond offering a theory which tries to understand how the mechanism of technological alignment works, the hope is also that this understanding may empower us to find ways and strategies to counteract certain developmental lines into the future—if that is what we want to do.

4.9 The rationale of responsible innovation

This section will not address the main issues or cover the extensive area of responsible innovation. The aim is simply to explore the affinities between this topical area and the analysis presented throughout this chapter.

Over the last 20 years, there has been a growing interest in responsible research and innovation (RRI) and so-called upstream engagement (Stilgoe et al., 2013; Wynne, 2002). Rather than focusing solely on the ethical assessment of products that are already invented and offered ('Responsibility of governance'), different scholars have emphasized the importance of what happens before the public typically gets a say about technology (Owen et al., 2021, 2). Is it too late for the public to discuss the ethical challenges of, for example, GMOs, geoengineering strategies or nanorobotics once they are on offer? Is the ethical and political debate merely a necessary step with no real impact, one that developers must endure, and which, if nothing else, continues to provide jobs for social scientists and philosophers?

Responsible innovation concerns the relationship between science, technology and society, and one of the major underlying issues within this field is the question of the autonomy of science (Stilgoe et al., 2013). The traditional division of labor between science and society is based on the is/ought distinction, where science deals with 'is' and society deals with 'ought' (see also Section 2.5). The classical idea posits that science plays an independent role and therefore bears no responsibility for the social, health and environmental effects of scientific endeavors. According to this well-known narrative, governments or other societal institutions should not dictate or pressure scientists or scientific institutions regarding research questions, outputs or methodologies; similarly, scientists should not engage in normative issues *as* scientists.

However, various accounts, especially within the field of Science and Technology Studies (STS), argue that this strict division of labor is either unsettled or has become problematic (Funtowicz & Ravetz, 1993; Gibbons et al., 1994; Lie, 2016, Chapter 1.4.1; Wynne, 1992). This perspective reflects the fact that science has become an essential and unavoidable part of modern capitalist societies through the "new" knowledge economy, which is perhaps the most significant factor spurring the RRI debate. A related but distinct aspect that has made this question urgent is the growing body of contested and controversial technologies that have emerged over the last 30 years (Owen et al., 2021, 2–3), a topic touched upon already in Section 1.2. As these developments occur 'on the ground,' we should not be surprised or offended by those who challenge the clear-cut division of labor between science and society, which is precisely what the discourse on RRI aims to address. When we observe that the most prestigious areas of science are linked to fields where the application of that knowledge can be profitable (techno-science), it is sometimes tempting to conclude that the purported autonomy of science can also be put to use for the wrong purposes.

It is, however, understandable and reasonable for scientists to express their views when they are uncomfortable engaging in debates about how societies should be formed. The fact that technology and science increasingly shape our societies does not imply that scientists should participate in those specific debates *as* scientists. I believe that the responsibility of scientists does not first and foremost concern the use of scientific results but a better understanding of the role science plays in our modern societies.

First and foremost, they should not assume that their scientific endeavors—followed by innovation, at least the innovation we observe today—are necessarily emancipatory (Stilgoe et al., 2013, p. 1568). Technological development may lead to emancipation in the sense that it allows us to perform the same tasks more (nexus) efficiently and broadens our capabilities. However, it does not facilitate emancipation if we examine the nature of the relationship between humans and technology.

I have argued that the alignment with technology is neutral. This simply means that a technology would be truly better than another if it is 'ready-to-hand friendly.' As we have seen, this is rarely the case. Ready-to-hand comportment does not depend on the technology itself; almost any technology can serve as the manifesting partner that enables ready-to-hand comportment. When we claim that one technology is better than another, we mean that it is relatively more (nexus) efficient. However, greater efficiency cannot equate to greater emancipation if efficiency is relative. I cannot say that I have 'moved forward' or become less enveloped if there is no endpoint toward which I am moving (cf. *sensitivity*). I cannot alleviate my hunger by consuming one more hamburger if it is impossible for me to achieve psychological or physical satiety. If alignment defines the meaning of technological use, one might argue *that the type of* technological development that routinely and inevitably renders our current use of technology present-at-hand *is less* emancipatory than, to take the other extreme, no technological development at all.

This is not an argument against technological development. It is rather a critique of the idea that science and technology inherently lead to emancipation and the notion that they intrinsically contribute to the quest for human betterment and liberation. What we seek in our relationship with the world (in the Heideggerian sense) is a ready-to-hand relationship that demands skillful subordination. If this sort of alignment is the essence of the human-technology relationship, it is also a fact that, at a certain point, the overwhelming influx of new technologies can lead to a more or less continous misalignment, rendering it no longer liberating.

This means that today's heroes—scientists and innovators—cannot *assume* that whatever they do is for the greater good. Science and technology generate possibilities, and new possibilities are not inherently good. Scientists and engineers should be aware of the mechanism of realignment through expansion, and there should be *no excuse* for those who naively assume they are doing good by default. Hiding behind the is/ought distinction in this respect should be questioned. The privileged position derived from this distinction must be justified in each case, rather than taken for granted.

Secondly, where there is no reconciliation, only expansion, there needs to be a realistic and competitive theory that can argue that our 'unsettled business with nature' can be resolved through technology. Continuous upheaval makes little sense if it does not lead us to 'a final restful place'—a reconciliation once and for all, an atheist paradise. A relationship with technology where the use demands nothing from its user, devoid of friction, frustration, hassle or effort. A situation where all external resistance has ceased. Is this possible?[12] To the extent that my claims about the role of sensitivity are correct, this dream appears to be a contradictory quest. The less resistance, the more sensitivity. The gateway to emancipation requires effort, as this is what a ready-to-hand relationship demands. If there is no realism in 'the final reconciliation,' we should conclude that today's technological development traps us in a false promise—condemning us to live in fractured, present-at-hand relationships with our manifesting partners. It's hard to see how 'greater efficiency' justifies the cost of such a predicament.

Without a deeper critique of our technological development, it is also challenging to present a positive alternative—an alternative that fosters development on the 'human side' of the equation. If we do not achieve what we desire as promised by technological development, it is more fruitful to seek advancements that cultivate even deeper ready-to-hand relationships.

An example from my region could be that during winter, a significant portion of the population in Tromsø ventures into the mountains on skis (randonnée) or snowboards (split-boards). Over the last 10–15 years, many individuals have begun changing their gear every year or every other year. The kind of development I am discussing would prompt the question: What do I need to do to reconcile with the pair I already have? With this question, I want to remind the reader of the following: We do not necessarily have to choose to act in a ready-to-hand mode. It occurs more or less automatically as part of a learning process. However, there are ways to transcend a "normal level" of ready-to-handness, which is significant in this context. On this topic, insights may be found in Zen Buddhism, pedagogical theories in sports, different understandings of flow, Albert Borgmann's theories (cf. focal practices), and Dreyfus and Dreyfus's models of skill acquisition. If this were our focal point, I am confident that humanity would devise new, inventive ways to align and reconcile with our manifesting partners. Perhaps there are "superpowers" we have yet to understand or investigate properly—something far more admirable and culturally enriching than what we can buy for money? (Something that could conquer transhumanist dreams of superhumans?)

If we recognize that 'belief in the gear' is mostly a false gate-opener to alignment, there is also the potential for seriously considering reconciliatory development. In some ways, this could also redefine what we mean by *responsible* innovation.

I have argued above that new technologies come into place whenever a new technology makes the current one conspicuous *and instantly provides the answer to the problem that the (new) technology has pointed out in the old* one. This mechanism is such a strong mechanism because the use of technology crucially depends on ready-to-hand use. When this use is broken by the new technology, it requires significantly more effort to reconcile than to expand, given the available options and capabilities. The most critical steps toward implementation have already occurred when the 'new *zeug*' is invented. This is why ethical and political debates following the invention are often rather impotent and why responsible innovation is rather important. More importantly, however, scientists and innovators cannot rely on the basic assumption that their efforts will generally lead to reasonable assessment, emancipation and human betterment. This places an even greater responsibility on those who participate in offering new possibilities, ensuring that those products are at least containers or components of containers that are as symmetric as possible (Section 3.3.5). If persuasion (or even compulsion) is already established by the ready-to-hand/present-at-hand dynamic, and democratic powers to control technological development are similarly limited, then enormous responsibility should rest on the shoulders of those engaged in technological development.

We have witnessed tremendous changes in our societies over the last 100 years due to technological advancements. Needless to say, my arguments here support those who believe that our reliance on politics and ethics to regulate technology 'after the fact' has been a complete failure.[13] With the account provided here, I hope to have offered a rationale for why RRI should be more than a mere buzzword that only makes a difference in the application for funding (Bensaude-Vincent, 2014). Additionally, this account presents an argument for why the public is justified in wanting to engage upstream, rather than at the end, whether as part of hearings, focus groups or as consumers (cf., e.g., post-normal science).

Notes

1 Heidegger is famous for his "construction" of a new language, which he justifies as an attempt to overcome some basic beliefs in the philosophical tradition that he thinks are erroneous. I will try to avoid an overuse of Heidegger's terminology here and use his concepts where it does not seem to stand in the way of our understanding. I will use the German expression *dasein* since the English translations are confusing, and the expression 'humans' is misleading in the context.

2 Humeans and modern physics have claimed otherwise. See, e.g., Ladyman and Ross (2007). As previously mentioned, some philosophers like Schelling, Whitehead and Bergson are typical exceptions in that they focused on the processual character of being.

3 Usually, we would call this condition flow, and one may assume that this is what the Zen Buddhist tries to achieve through spiritual praxis. That is to say that it would be an exaggeration in the opposite direction to claim that this is the normal state-of-affairs for our ready-to-hand relatedness.

4 I state that this is not the topic here, but the reader would probably recognize that Heidegger's account differs from the account given in Chapter 3, where I clearly avoid a phenomenological or Heideggerian perspective on ontology. See also Section 2.10 and Lie (2025) where I clearly argue against the epistemological dogma, which in this context means that I argue that we do not have to accept either the traditions scientific subject–object stance or the phenomenological stance as the only alternatives available.

5 Instrumental and goal-directed rationality are sometimes conflated. Max Weber distinguished between goal-directed and value-directed rationality. Instrumental rationality represents a third type, where efficiency becomes the primary goal. This distinction was particularly emphasized by Jacques Ellul (1964, XXV).

6 The technological imperative, or the gravity of technology, states: "What is possible shall be realized." This principle suggests that ethical, political or other considerations often lose significance when confronted with the sheer possibilities (efficiency) that technology presents.

7 One may compare this with what Charles Taylor refers to as weak evaluations.

8 One may speculate whether Heidegger would agree with the distinction between basic and nonbasic needs, or more specifically, with Maslow's hierarchy of needs. In *Being and Time*, he addresses the role of desire (*Wunsch*) as part of the discussion on authenticity (239). Generally, however, 'identity' is more primordial than basic needs, to express it in non-Heideggerian terms. *Dasein* is not defined by fixed needs; it must continuously decide 'who I am and who I must be' (even though this question is often obscured by habituation). Heidegger would argue that our most pressing need is to 'serve' *jemeinigkeit*. That is, we are willing to starve and go to war to defend *who we* are and so on.

9 This might serve as a curious example: In Norway, we have something called '*dugnad*.' A *dugnad* occurs when you enlist neighbors and friends to help with a practical task requiring "man-power," usually with food and drinks provided in return. On such occasions, there is often one or two individuals who struggle to engage (ready-to-hand) because they are constantly and "helplessly" occupied by suggesting 'better solutions.'

10 Harari (2014) and others have argued that religion plays an important role in organizing people to overcome certain limits. While this seems plausible, it is debatable whether religious institutions should be functionally explained in this way (Elster, 1980).

11 Opposite views to the account given on 'possible at all' here might be for instance the previously mentioned idea of our transcending Promethean or Faustian nature, or the so-called voluntarist view, alluded to throughout this chapter, which would not only analyze technological development from an individualist point of view, but also interpret the development as a result of *rational* action based on conscious and careful consideration.

12 Some literally believe this is possible. For example, Ray Kurzweil who thinks this is the final endpoint of singularity, i.e., the technological development (see, e.g., Epoch 6 in Kurzweil, 2005).

13 Habermas (1984) would be the appropriate reference here. His discourse ethics is based on the premise that we can guard against instrumental rationality by maintaining a strict division between science and technology and the discursive realm of ethics and politics. His theory assumes that the mechanisms underlying technological development can be tamed by discursive rationality.

References

Bensaude-Vincent, B. (2014). The politics of buzzwords at the interface of technoscience, market, and society: The case of public engagement in science. *Public Understanding of Science*, *23*(3), 238–253. https://doi.org/10.1177/0963662513515371

Dreyfus, H. L., & Dreyfus, S. E. (1986). *Mind over machine: The power of human intuition and expertise in the era of the computer*. Free Press.

Ellul, J. (1964). *The technological society*. Alfred A. Knopf.

Elster, J. (1980). Cohen on Marx's theory of history. *Political Studies*, *28*(1), 121–128. https://doi.org/10.1111/j.1467-9248.1980.tb01239.x Digital Object Identifier (DOI)

Funtowicz, S., & Ravetz, R. (1993). Science for the post-normal age. *Futures*, *25*(7), 739–755. https://doi.org/10.1016/0016-3287(93)90022-L

Gibbons, M., Limoges, C., Nowotny, H., Schwartzman, S., Scott, P., & Trow, M. (1994). *The new production of knowledge: The dynamics of science and research in contemporary societies*. Sage Publications.

Habermas, J. (1984). *The theory of communicative action: Reason and the rationalization of society*. Trans. T. McCarthy. Beacon Press. (Original work published 1981).

Harari, Y. N. (2014). *Sapiens: A brief history of humankind*. Harper.

Heidegger, M. (1977). The question concerning technology. In D. F. Krell (Ed.), *Basic writings* (pp. 311–341). Harper & Row. (Original work published 1954).

Heidegger, M. (1992). *Being and time*. Trans. J. Macquarrie & E. Robinson. Blackwell Publishers. (Original work published 1927).

Horkheimer, M., & Adorno, T. W. (1972). *The dialectic of enlightenment*. Trans. J. Cumming. Herder and Herder. (Original work published 1947).

Inglehart, R. (1977). *The silent revolution: Changing values and political styles among Western publics*. Princeton University Press.

Kurzweil, Ray. (2005). *The singularity is near: When humans transcend biology*. New York: Viking.

Ladyman, J., & Ross, D. (2007). *Everything must go: Metaphysics naturalized*. Oxford University Press.

Lie, S. A. N. (2016). *Philosophy of nature: Rethinking naturalness*. Routledge.

Lie, S. A. N. (2025). Making sense of 'description from within': An ontological challenge to (eco-) phenomenology. In M. T. Heneise, et al. (Eds.), *The Routledge handbook of ecophenomenology*. Routledge.

Lundvall, B. Å. (1992). *National systems of innovation: Towards a theory of innovation and interactive learning*. Pinter Publishers.

Marcuse, H. (1964). *One-dimensional man: studies in ideology of advanced industrial society*. Beacon Press.

Owen, R., von Schomberg, R., & Macnaghten, P. (2021). An unfinished journey: Reflections on a decade of responsible innovation. *Journal of Responsible Innovation*, *8*(2), 217–233. https://doi.org/10.1080/23299460.2021.1948789
Stilgoe, J., Owen, R., & Macnaghten, P. (2013). Developing a framework for responsible innovation. *Research Policy*, *42*(9), 1568–1580. https://doi.org/10.1016/j.respol.2013.05.008
Weis, T. (2010). The accelerating biophysical contradictions of industrial capitalist agriculture. *Journal of Agrarian Change*, *10*(3), 315–341. https://doi.org/10.1111/j.1471-0366.2010.00273.x
Winner, L. (1977). *Autonomous technology: Technics-out-of-control as a theme in political thought*. MIT Press.
Wynne, B. (1992). *Carving out science (and politics) in the regulatory jungle. Social Studies of Science*, *22*(4), 745–758. https://doi.org/10.1088/0963-6625/1/3/004
Wynne, B. (2002). Risk and environment as legitimatory discourses of technology: Reflexivity inside out? *Current Sociology*, *50*(3), 459–477. https://doi.org/10.1177/0011392102050003010

5 Freedom and limits

In this chapter, I primarily discuss the concept of freedom in the context of relation-limiting cosmology. This focus means that I don't engage, more than necessary, with classical questions of freedom. These classical questions are classical exactly for the reason that they arise within the framework of part-limiting cosmology, where issues such as determinism, freedom of the will versus freedom of action, positive versus negative freedom, the role of rationality and social influence make sense. Certain ontological assumptions that are often taken for granted, make it difficult, if not impossible, to reconsider the relationship between our notion of 'freedom' and limits in nature. As we already saw in Chapter 2, a relational view of freedom is not merely a modification or addition to our established perspectives on freedom.

At this point, however, it is not completely clear to me what a relational view of freedom entails, and I do not believe I have the necessary capacity and oversight to understand and work out a relational view on freedom as such. Therefore, I can only explore where some basic assumptions lead and see if it is possible to develop a fairly consistent perspective that specifically addresses the issue of freedom and limits. First, however, we need to recollect what a part-limiting view of freedom and limits entails.

5.1 The prevailing concept of freedom

In light of the discussion in Sections 2.4 and 2.5, we can suggest that what we commonly understand today as 'freedom' is an invention that arose from a part-limiting cosmology. In this cosmology, goal-directed behavior is imagined as originating within the individual agent—goals are internally formulated, and the world is perceived as a collection of separate, unentangled parts that are connected only by external causal influences, if connected at all. The independence or autonomy of the agent (freedom of will) is the precondition that makes it possible for the agent to navigate this world of external obstacles and opportunities and which makes it possible for the ideal agent to choose both ends and means rationally and deliberately. This is what it takes to exist and maintain oneself as a separate, independent part, that is, as someone who has the capacity to act freely.

DOI: 10.4324/9781003307372-6

In contrast, within a whole-limiting ontology, goals are not strictly located "inside" the agent. Instead, they are woven into an integrated structure in which individual aims are meaningful only insofar as they connect to the broader purposes of the whole. What appears as autonomous goal-setting in the modern part-limiting view would, from the whole-limiting perspective, be understood as embedded participation in a larger unfolding order.

As we no longer see ourselves as a part of a meaningful and morally good order, it follows that to realize our part-limiting freedom, we must do so in competition with other parts. This is not seen as inherently problematic—unless those parts are other humans. When other humans are involved, the situation becomes more complex, requiring a *chosen* state order to regulate our interactions, often referred to as democracy. In contrast, with allohuman parts, we rely on science and engineering. These tools not only help us achieve our goals more efficiently but also enable the pursuit of previously unimagined possibilities, thereby enlarging our capacity to act freely. Thus, the pursuit of freedom would be achieved where we have a state that optimally regulates our freedom concerning other people and where our instrumental goals—restricted by other allohuman parts—are accomplished through the technoscientific manipulation of these parts.

This view has, of course, received various criticisms throughout the centuries. I will briefly mention some of them here, although some of them have already been mentioned in other contexts.

One criticism concerns the agent that holds the goal. In what way and to what degree can we say that this agent 'is a self"? 'Is the self a physical or a non-physical entity' and 'Is the self a given entity, or is it always under construction?' (existentialism); 'Can we distinguish between an authentic self and an inauthentic self?'

A related criticism questions our assumed goal-directedness. Are we goal-directed by virtue of our rationality or our feelings? To what extent is enlightenment a precondition for a type of goal-directedness that is free? Do our goals truly originate from subconscious drives (Freud)? Are our goals already embedded in a pre-established social and functional order (Hegel and Heidegger)? Are our intentionalities embedded in our bodily being, which extends far beyond our cognitive self (Merleau-Ponty)?

A somewhat different criticism asks about the nature of our goals. Is it true that we are guided primarily by what we regard as the highest values in life? And what are those values, by the way? Is it possible to agree on those values, or must we remain neutral regarding others' choices? To what degree can and should we distinguish between means and goals? The Frankfurt School, with Adorno and Horkheimer as forerunners, questioned the very core of the Enlightenment project when they claimed that the rationality supposed to guide us toward honorable goals had morphed into a calculating rationality where the goal is simply 'more efficiency.' A more subtle critique asks whether the goals truly lie in the means; that is, what is honorable does not lie in the goal itself, but in *how* you achieve them (e.g., Zen Buddhism).

A fourth kind of criticism questions whether it is true that all nonhuman creatures are not goal-directed. This criticism mainly emerged from the environmental and animal liberation movement in the 1960s and has achieved a broad theoretical base (see Section 6.5).

A related question has sparked a fierce and ongoing debate about what used to be called sociobiology. Ever since Richard Dawkins claimed in 1976 that all organisms, including humans, are merely 'vehicles for their genes'—with reproduction and survival as the only goals that matter—there has been deep disagreement about how we should understand ourselves as biological beings.[1] To what extent are we limited by our own biology?

A long-standing but increasingly prominent line of critique challenges the passivist ontology itself. This critique draws from diverse sources, including certain interpretations of quantum physics, the process philosophies of Spinoza, Schelling, Whitehead and Bergson, feminist theory, dispositional theory and realist approaches that reject the primacy of epistemology (i.e., New Materialism). We might also mention panpsychism, which emerges from efforts to address the so-called hard problem in the philosophy of mind.

I would argue that all of these criticisms occur within the framework of a part-limiting cosmology, except for the last type of critique. The first type of critique highlights problems within the already accepted definitions of freedom. For instance, the issues of rational choice or determinism question whether we can truly act freely, rather than questioning the very concept of freedom itself. To examine the concept itself, we must return to our basic understanding of nature (i.e., the last type of criticism). To the extent that we frame criticism in terms of ontology, where realist alternatives are also presented, I consider this promising. The important questions raised by the part-limiting cosmology will not have the same weight within a relation-limiting cosmology. The question of autonomy, for instance, is a question that *originates* from thinking about other parts of nature as separate and heteronomically determined.

An important challenge is that the only apparent alternative to the part-limiting discourse seems to be the whole-limiting discourse, which very few of us, including most critics of the ontological core of part-limiting cosmology, are willing to accept. For many, a whole-limiting view would translate into some form of illiberal order. If we view limits in nature as stemming from an integrated order, alarm bells begin to ring. In other words, our current understanding of freedom inhibits us from exploring alternatives to part-limiting frameworks, as the only alternative appears to lead us back toward a whole-limiting cosmology. This leaves us in a position where the only limits to our freedom that encompass all differences are those that are physically impossible to transcend, making all other types of limits contingent.

So far, I have proposed that limits are relational and more or less specifically integrated by the natural history of our universe. These limits constitute a certain order. I have not claimed that this order has a meaning or moral status. I have only stated that we need to understand our freedom in light of this order. If we fail to do this, we risk ending up in a human created but nonchosen whole-limiting order as a consequence of pursuing part-limiting freedom. This is at least what I will argue in the last parts of this chapter (see Section 5.5 in particular).

5.2 The ontology of freedom

Despite my insistence on the connection between freedom and ontology, I believe we should take a step back and start this section by asking: Is it even viable to examine 'the ontology of freedom'? Aren't freedom and ontology two categorically different matters, where you can have all sorts of ontologies going along with all sorts of theories of freedom?

I would like to briefly remind the reader of the reasons for rejecting this view. In addition to the assertion that our concept of freedom is rooted in a part-limiting ontology, we should recall that the very idea of autonomy is considered the pivotal mark of freedom because the passivist ontology is accepted in the first place.[2] Not only are parts in our universe seen as determined by antecedent causes, but they are also passive. Autonomy is the *ability* to transcend the causal determinism of nature (antecedent causes) to which we would otherwise be passively subjected. (In Kant's thinking, this comes down to the ability to give ourselves laws.) The more humans can withstand and overcome nature's inclinations, the freer and more human we become. To think and act as if ontology and freedom are independent matters is, therefore, to accept the ontology currently supporting our idea of freedom. Our conception of freedom did not originate from an investigation of *freedom* as such (Chapter 2). This becomes especially evident when we consider how adopting a relational ontology of nature, rather than a passivist one, reshapes the conceptual landscape.

Instead of believing that we possess our properties independently of "everything else," a relationalist asserts that humans, like everything else, are constituted through the relational nature of their dispositions. A relationalist would argue that even properties we consider to be specifically 'intrinsically human' exist and arise in relation to other parts. Language, feelings, consciousness, perception and rationality are all intentional properties; they are *about* something. Our more general properties, such as the digestive system, relate to what is digestible; our breathing relates to the specific composition of the air surrounding us; our hands relate to what is graspable, and so on. All our dispositions extend beyond ourselves and depend on something 'external.' Obsessed with the independence of our will, we conflate 'dependent on' with 'determined by,' viewing relatedness as a threat to our freedom. 'I depend on my biology' becomes 'I am determined by my biology.' When we perceive our biology as a threat, we typically defend our freedom by either attempting to demonstrate that our mind operates independently of our biology (e.g., the neuronal functioning of our brain) or arguing that there is 'a certain freedom' inherent in biology (e.g., by deploying quantum field theory to describe the workings of our brain). In the first instance, we claim that we are free because the determinism of the world does not affect us; in the other, we argue that we are free because the world (in this case, the biological realm) is not determined. In both cases, relatedness and dependency are construed as forms of external agency, which, under a passivist account, reduce humans to passive objects—a view fundamentally at odds with what is then defined as freedom.

When we consider the ontology of freedom from a relational perspective, we are not preoccupied with questions of determinism or whether freedom requires that our mind is 'an extra ingredient in nature,' as these take certain ontological claims about parts and relations for granted. From a part-limiting perspective, a relational ontology would threaten our independent and intrinsic nature, which would rule out the possibility of freedom altogether.

The concern that 'a relational theory might rule out the very possibility of freedom' may be justified if relationalism implies that there are no (real) parts (i.e., if the relata disappear into the relation). However, this is not the case with relational realism, where the '*modified* neutral monist view' safeguards the reality of parts (Sections 3.2.3 and 3.2.4). In the case of organisms, this involves dynamic processes that continuously manifest homeostatic functions, securing the organisms' *relational* independence and enabling some form of goal-directed behavior.

Relational independence means that the independence of an organism depends on its relations (as any evolutionary theory would assert). The fact that an organism has homeostatic mechanisms making it relatively independent when it needs to 'absorb water,' for instance, does not mean it is not dependent on (the relation to) water. There is a distinction between relationships and relations. I may 'end my relationship' with my loved one, but I cannot escape the *relation* that this relationship instantiates. This 'love-relation' carries certain characteristics that impose general demands on me—ethical, emotional or existential. If I refuse these demands, I am not merely leaving my partner for the wrong reasons; I am attempting to reject the relation itself. The fact that we cannot escape our dependencies does not mean I cannot escape the instantiation or the token of the relation, but rather that I cannot escape the relation itself. In a part-limiting view, I exist among other parts that *eventually* enter into relations. In relation-limiting views, these parts are instantiations of already existing relations. I may treat these parts as random and contingent, but that does not eliminate the relation.

Moreover, in most cases where I engage in activities beyond those strictly necessary for maintaining my existence, I still enter realms of nonchosen relations. If I want to grow potatoes (which I don't need to do), I must weed the land, but that does not eliminate the relation. I may hire a machine or a person to weed the land, but the relation remains; it is simply represented differently (Section 3.3.5).

From a relational perspective, freedom of action becomes a matter of how we relate to our dependencies, rather than a question of which relations I (freely) want to establish and 'depend' on. This means I am not *obsessed* with the *independent* nature of my partners (i.e., whether they possess rational agency or are merely passive contingent beings) because the dispositional agency of my partners is relational. Instead, I would be focused on what the relation entails, because that is what ultimately limits what I can do.

Let's assume that humans are 'social' beings. What does that entail? Perhaps something akin to what Aristotle suggested in his *Nicomachean Ethics*? Even if we don't fully agree with his view, we can still see the virtues he suggests as examples of what 'social' means. Being generous, courageous, reliable and trustworthy are possible responses to this relation. On a relational account, it would therefore be the relation that "defines" our freedom. To act in spite of, or ignorance of the relation, will not lead to freedom, as such actions will defy your ends.

As individuals, society and culture, we can misunderstand various relations. Thus, to search for freedom would mean seeking the right understanding of what the relation entails. Such an understanding would involve a praxis where your actions do not undermine the 'relational demands.' Hence, 'what the relation entails' would also roughly describe the limits of your freedom: If you want to act freely, this is what makes it possible. For instance, if you wish to act freely in a particular social situation, you may depend on your previous cultivation of generosity, empathy, curiosity, etc.

To be able to 'relate (adequately) to our relations' is not a genuine human disposition. When a fox is curious, it is interested in how to relate to the part(s) *that it already has a relation to*.[3] Evolutionary adaptations are patterns of behavior which are probing answers to how a particular part 'should' relate.

According to relational realism, all parts are dispositional *partners* within a system of already established "leanings" (cf. possible-N). When we aim to manifest something—that is, to achieve a goal—we always do so in conjunction with other parts. All manifestations are mutual, which in the case of humans translates to: We never act alone. *We are mutually manifesting partners*. Reaching for a goal does not require cooperation as co-operation is already a part of acting. The fact that we are goal-directed does not imply that relations become or should become contingent. If the other parts in case F are merely contingently associated (with each other and with me), there is no relation to relate to. They are simply 'parts.' Hence, if I relate to the other parts in case G as if they are only contingently connected, while in reality they are not, the precondition for freedom is compromised. You cannot achieve the goal of becoming a friend if you believe that the relation (friendship) is not bound by certain demands. This is why the ontology of freedom matters.

In some cases, you may even need to relinquish your goal when it conflicts with your dependencies. This notion aligns with Immanuel Kant's argument that certain actions are forbidden or self-defeating. That is: We may interpret the categorical imperative as a rule that indicates whether your action is (or is not) compatible with the dependency that makes your action possible in the first place (i.e., a relationship of trust is a precondition for the possibility of lying, which makes lying self-defeating). See also Section 6.5.2.

Freedom as striving for independence entails the belief that our ability to achieve our goals relies on our capacity to remove or pacify parts that hinder us. These are perceived as contingent limits to our goal-directed actions. Thus, limits that cannot be overcome in this sense stand out as *real* limits. This is why we consider it a breakthrough and a remarkable achievement when something we believed to be a limit (in this sense) turns out not to be one. Consequently, we repeatedly think we can transcend limits and get rid of dependencies through technological advancements without costs or repercussions. Therefore, assuming that physical limits (e.g., the speed of light, etc.) are the only *real* limits entails not only a particular view on what limits are but also a *particular* view of what freedom is.[4]

Freedom means to stay with the trouble (cf. Haraway, 2016). There is no escape. Whatever 'self we choose to maintain,' there will always be limits and dependencies. We are, as Schelling would say, 'made of limits.' Limits are inherent to our being, whatever we choose to be.

5.3 Fleedom

To assert that we conceptualize the world merely as a collection of contingently arranged components available for our disposal seems a bit overwrought and forced. It gives the impression that we are akin to immature psychopaths who have ontologically sanctioned ourselves to assume a god-like role to the detriment of everything else. While I contend that we indeed perceive certain elements as independent and contingently arranged (in the sense that they are *not* relationally organized), I do not believe this is a widely held *explicit* ontology. There is no reason to doubt that people today often demonstrate considerable concern for the environment and, for the most part, feel a connection to the fabric of the world. There exists a sense of awe, wonder and gratitude. In other words, there are specific contexts in which this perspective manifests.

As described in Chapter 4, I first and foremost experience that my will is entangled in a nexus of pre-existing dependencies. For instance, if I wish to engage in activity X, I must concurrently perform actions F, G and H. To cultivate potatoes, one must also weed and sometimes irrigate the soil. While we may undertake weeding, there are instances—particularly when a more advantageous opportunity arises (as discussed in Section 4.5)—where the relationship becomes conspicuous. Thus, I might opt for the herbicide, which alleviates the need for manual weeding. This choice allows me to pursue X without necessarily engaging in Y. However, this does not inherently imply that I view the relation as contingent (only the particular relationship), *but that is also the question*. Do I or do I not perceive the relation as contingent? In most instances, I believe we do, as we often transition to the next (more 'efficient') solution to what we consider a problem (e.g., weeding manually becomes present-at-hand) *without further reflection*, thereby assuming we have rendered ourselves independent of the relation. We neglect to consider, all factors considered, what would constitute the optimal limit arrangement. Instead, we adhere to what I term the 'template of fleedom,' which is: 'I want X but I don't want Y.' We seek to escape what constrains our actions and what we depend upon in the pursuit of our objectives. A psychopath may command: 'Eliminate that animal, procure that individual, remove those obstacles, and report to me when everything is *in place.*' Most of us remain silent, simply proceeding to purchase the weed killer.

Why is this template important at all? It is important because or to the extent that X and Y are relationally connected in the first place. Much of our behavior within Western consumer culture revolves around executing or fantasizing about scenarios wherein we can attain X without Y. The elimination of Y facilitates or simplifies the acquisition of X. For example, I desire to tan without suffering from sunburn (or skin cancer). I wish to consume meat without the necessity of slaughtering an animal. I aspire to snowboard downhill without putting in the labor of walking uphill. I yearn to become intoxicated while avoiding a hangover. Nature 'comes in package deals,' yet I wish to reject the entirety of the package. This phenomenon illustrates *how* relationships become contingent, and when questioned, our scholars might come in handy to justify our actions by referring to the passivist contingency ontologies of science.

A pertinent example of such scholars would be utilitarian 'thought-experiment-philosophers.' In 'The Fable of the Dragon-Tyrant' (Boström, 2005), Nick Bostrøm asks whether we would be willing to accept a situation where a dragon demands daily and more or less random offerings of human lives as a fact of life or whether we would or should 'fight the dragon.' The fable puts the reader in a situation where "science" obviously has made it possible to 'have life without death', which in itself is a striking example of the fleedom template. So, what would you choose? Death is no longer a condition for life. A similar example of cleverish thought-experiment-philosophy arguments I came across in a debate: 'If you don't like the idea of a life expectancy of 300 years because you fear a boring life, why don't you suggest that we cut our life expectancy to 30 years? That would saturate every moment with meaning!' Could we just play around with numbers like this and assume that everything else would be equal? Apparently yes. If you are worried about reproduction, we can also make children raise themselves after the age of 10. Push this button and let it happen. From this outlook we obviously live in a world of contingent relations and limits with randomly arranged parts that can be rearranged at our discretion.[5]

This process of disentanglement also propels economic growth. New limit arrangements manifest as products or services in the market, such as (in the cases mentioned above) sunscreen, butchers, ski lifts or aspirin. This encapsulates the essence of human expansion; we are enmeshed in dependencies from which we seek to extricate ourselves. Through the ongoing '*contingency* extension,' a continuously expanding array of means to achieve X without Y proliferates at an almost unfathomable scale.[6]

From this vantage point, development can be understood as the process of 'pulling things apart.' The facilitators of this process predominantly include the state, scientific advancement and technology. The most effective economic system for coordinating and enforcing this process is capitalism (or, at the very least, a market-based distribution system), which provides both goods and services. We collaborate (through the enabling possible-I systems) to transcend common dependencies, transforming them from shared burdens into individually chosen arrangements. When we criticize ourselves for this, we label it (toxic) individualism; however, more often than not, we celebrate the same process—framed through the 'I want X, but not Y' template—as a narrative of increasing freedom. I have proposed to refer to this process as 'fleedom.'

In seeking fleedom rather than freedom, we exhibit an unexamined tendency to assume that dependencies and limitations *dissipate* when we establish our own chosen limit arrangements. As seekers of fleedom, we maintain the belief that the underlying relation connecting X and Y ceases to exist when we "eliminate" or substitute the specific relationship between X and Y. We externalize the relationship by reducing it to mere elements or parts that obstruct our progress. We fail to recognize that, in disentangling, we become entangled in the altered dependencies of the same relation. The general driving force is the idea that we will end up in a good place when we eliminate or 'falsify' the obstacles.[7] True freedom, it is thought, exists where there is no resistance or external hindrance. By orchestrating our own package deal, we ultimately achieve freedom. Thus, fleedom practically exemplifies the instantiation of the passivist perspective, wherein all components are contingently arranged from the outset.

The rationale for distinguishing between freedom and fleedom, as articulated here, lacks coherence unless we assume the existence of a relational human identity, however. Within a part-limiting framework, such a distinction does not hold. Fleedom becomes indistinguishable from freedom when I perceive myself as an independent being, because I can only realize my identity by abolishing my dependencies. 'I become myself when my essence is not disturbed by external factors.' This perspective aligns with the part-limiting notion that positions itself *in opposition* to the whole-limiting concept. The part-limited drive toward fleedom is a *reaction* to what is seen as a threat from the whole-limiting trunk.[8] However, freedom devolves into fleedom when contested by a relation-limiting framework. Thus, the common thread of fleedom is the inclination to disregard the relational dimension of reality. More than a proactive quest for freedom (whatever that may entail), it embodies a pursuit of circumstances in which we *can* escape our relational dependencies as such. If the underlying rationale of fleedom was true, it must also be true that the liberation from specific dependencies does not create new ones.

I am not making the argument that we are incapable of occasionally anticipating certain challenges that may arise when we desire X but not Y; rather, I argue that we perceive such challenges as transient and not a fundamental aspect of reality itself. The underlying premise is that our dependencies are fundamentally contingent and that we inhabit a reality characterized as "loose and separate" rather than relational in nature.

5.4 Optimum and freedom

If limits are relational rather than external and separate, they must be about optimums. If our relatedness (as described in this book) is inescapable, there is no way for me to maximize my freedom infinitely. In certain instances, I can hypothetically become freer than I am now, but there will always be a limit—a ("theoretical") point at which I can achieve no more freedom. This limitation arises because any goal-directed behavior is a co-manifestation of relationally tied parts.

When we say that there is a limit to how many individuals of a certain species can inhabit a specific area, we are referring to the optimum population size, which is sometimes called the carrying capacity of the biotope. This does not mean that the population cannot exceed this number; rather, it cannot grow larger given the specific set of partners involved. An optimum, if the concept holds any meaning, is the 'balanced result' of the multiple partners involved.

The term 'optimum' can be interpreted in two ways: normatively or descriptively. The most common usage of optimum is normative. Something is optimal for a certain agent, meaning it is 'the best' or 'the perfect' for someone or something. Reaching an optimum equates to reaching a limit—the best possible state—where something cannot be better than the best. Using optimum as a descriptive concept is more problematic, as it would require some form of general teleology in nature. In this case, the optimum would not be for a specific individual or entity, but for the whole (system). For homeostatic systems like our body, we can accurately discuss and describe an optimal level of sugar in our bloodstream or an optimal level of calcium in our bones, but this is only because organisms already exhibit teleological behavior. However, as previously argued, it is difficult to assert that nature itself displays teleological behavior. Therefore, since we are interested in freedom, we must concentrate on normative optimums.

If we understand nature within a part-limiting framework, we can create a situation that is (more or less) optimal for our goals by assembling (discrete) parts in a way that supports those goals. However, if we view nature through a relation-limit framework, any such part-limiting optimum would be *trivial*. The partners we combine (the new limit arrangement) are not more optimal; they are simply different.

To act freely (or not) involves some form of goal-seeking behavior, although this does not always manifest as 'representational intent.' At a minimum, there must be some kind of wish or urge involved. If this wish is acted upon and not stalled, it produces goal-seeking behavior. In our goal-seeking behavior, the best we can hope for is to reach an optimum. When something has reached its optimum, it has attained *the limit* for the specific goal-directed activity in question. Reaching this optimum signifies that you are as free as you can be. The phrase 'can be' is crucial here. From a relational perspective, it makes sense to understand the optimum as the "balanced" outcome of all the manifesting partners involved, but this understanding leaves us with a crux. If the optimum is indeed the "balanced" result of all the manifesting partners, then the concept of optimum seems to be a trivial or even a "deterministic" fact.

For instance, you could have run faster *if* there had been no headwind. However, since there was a headwind, you could not have run faster. If you wish to run faster, this would not have been possible due to the manifesting partners involved. You could not have reached a higher optimum, even though you desire to do so. One way to run faster would be to fend off the headwind, perhaps with a physical barrier. You could also run faster with the same parts if you did more training, but the optimum would still be trivial, as the manifestation is merely the result of a changed partner (similar to how a physical barrier stops the headwind). In other words, you could adjust the manifesting partners involved (including yourself). This would enable you to run faster, but the optimum would still be trivial. This means you can run *faster* by rearranging the other parts, yet you would never exceed a trivial optimum. The reason is that, from a part-limiting perspective, the parts are nonrelational. This illustrates that freedom and optimum are not about optimizing partners but about optimizing *within* the realm of the partners. To reach for an optimum is to seek the most effective mode of collaboration between partners—essentially, to realize the latent potential in mutual manifestations.

While this may initially seem overly idealistic or vaguely normative, the relational realist view grounds this process in concrete dispositional interactions. At this point, we need to recall that it is not abstract dispositions, but concrete *bearers* that come together in a manifestation (cf. Section 3.2.5). The part-limiting view describes the congregation of parts as a gathering of fixed properties or partners that contribute their essential attributes. A situation that might be best accounted for by a vector model (Anjum & Mumford, 2011, Chapter 2). In contrast, the relational realist view portrays this gathering as a congregation of partners (bearers) with multiple properties that contribute dynamically (Section 3.2.5). Each partner has a set of potential contributions that depend on the other partners. While *the set of partners* might be fixed, their contributions are not; one partner can influence the contributions of others.[9] This reciprocal agency is the leeway that an agent, such as a human being, possesses to reach something more than trivial optimums. This influence or agency may occur by chance, but what if we want to do this intentionally? By changing my contribution, I can change the contributions of the other partners… Could this be achieved by gaining a systematic oversight over all potential contributions from each partner to calculate a set of procedural actions? I don't think so.

The only—or at least the most efficient—way for the optimum to become something more than trivial is through subordination. To achieve a synergistic optimum, one must be a part among parts, rather than an outsider to the partnership.

Consider the simple task of chopping wood: it requires, at a minimum (in addition to the human agent), a set of relational partners such as the axe, the earth (gravitational pull) and the piece of wood itself. All these bearers have certain *reciprocal* dispositions that the human partner must take into account. The key question is: how is this consideration achieved? Since the partnership is reciprocal and changes dynamically, the human agent cannot view these partners from the outside to reach the optimum. The other partners have their own "agency," and to utilize this agency optimally, the agent needs to engage. It must find a balance between listening and acting. Similar things can be said about a social gathering.

In Chapter 4, we examined Heidegger's claim that the human agent (Dasein) must subordinate itself to the relational field of 'significance.' The agent cannot freely position itself outside the interactions with other manifesting partners, as they are not separate entities functioning as "essential objects." The human partner is not free to act solely according to personal preference but must *engage* in the intra-actional field that opens up a navigational horizon for optimal courses of action. According to this account, we are not in a situation where I have set a goal and where other entities can be categorized as supportive or obstructive to my freedom. Attempting to identify supportive and obstructive partners does not make sense, as "support" and "obstruction" are intertwined and arise dynamically from the intra-action among partners, not from the single parts.

Various teachings and perspectives address the phenomenon of subordination. In the Western tradition, Romantic philosophers idealized the artist (the genius) as one who can dissolve into the whole and produce works of art that express the inner truth of nature or spirit, as if channeling a deeper unity beyond the self. And as we have seen, Heidegger discusses the concept of ready-to-hand (*zu-handen*) concerning tools and later introduces the idea of releasement (*Gelassenheit*) in relation to beings in general. I have also referenced a research-based account, grounded in Heidegger's '*zu-handenheit*,' developed by Dreyfus & Dreyfus in the 1980s, as well as Albert Borgmann's ideas about the prominent role that focal practices play. More recently, Hartmut Rosa has proposed a theory of *resonance* that has a sociological flavor but importantly does not exclude our relationships with allohuman parts. The most enduring and comprehensive 'school of subordination' may be found in the Zen-Buddhist tradition, where humans, *as* manifesting partners, are trained to embody 'the manifestations of the relational whole' (Phillips, 1962, xvii).

Thus, the assertion here is that you are, more or less, forced to subordinate if you want to achieve a relational optimum. You must subordinate to the 'will' (dispositions) of your partners to attain what you desire. You need to let the axe 'run its course' to split the log optimally. When your goal is immersed in the activity of 'reciprocal forces,' there is also a demand for subordination. Typically, we refer to this as a skill. Skillful performance is how relational limits are *enacted* with respect to a goal (optimum). Skill is not about becoming adept as an isolated entity, but as a collaborative participant.

Pursuing fleedom, on the other hand, suggests a conflict between subordination and independence, where the conflict is resolved by stacking: adding or subtracting parts to achieve an 'optimum.' Fleedom leads our life into consumerism, which always leaves us unsatisfied. A global trade of goods is of no help. Fleedom seeks ideal partners, while (relational) freedom aims to optimize within the already existing intra-actional field. The result of fleedom is trivial optimums, which are genuinely limitless. With trivial optimums, you can continuously become 'freer,' but what you truly seek is a limit—to be as free as possible—which is only attainable in nontrivial optimums.

Subordination demands effort that leads to skill. Refusing to subordinate makes your partners conspicuous (present-at-hand), leading you to 'put the blame on the partners.' This blame persists in endless forms whenever you deny the opportunity to become an 'insider.' In fleedom, we cast ourselves as "objects" by blaming our partners. We end up as "objects" among objects stacked as a trivial partner in a trivial optimum. We insist on being rigid like the supposed part-limited object, evading the opportunity to reach for the optimum—the limit of what's possible (given partners x, y, z, q, r). True freedom lies in the power to subordinate, to create something beyond trivial optimums.

At the center of fleedom lies the idea that my freedom is in conflict with my manifesting partners. They are not viewed as partners in manifestation but rather as contingent parts in a world of separate entities. The belief is that by shaping these parts to align with my will, I will become freer. However, as previously noted, this is an endless endeavor, since relationships, challenges and dependencies will never disappear. Attempting to escape entanglement often leads to re-entanglement occurring *behind my back*. Fleedom tends to be grounded in the dim notion that we are moving toward a state where dependencies are drastically reduced, eliminated or at least entirely chosen—a condition in which mere intention is enough to carry the will through. But this ideal falters, as our *sensitivity* tends to increase more or less in proportion to the diminishing resistance (Section 4.7). Thus, it is not arguments, but fleedom itself that advocates for perpetual economic growth. More importantly, resistance does not vanish; it merely reshapes into new configurations. This will be the topic of the next sections.

5.5 Optimization and the production of a possible-I order

In Chapter 3, I pursued the idea that limits in nature extend beyond the fixed constraints found in physics. I argued that what possible-N *nature does or has the capacity to do* should serve as our starting point for understanding limits in nature. What possible-N nature does include limits in physics but also those in biology and ecology, among other fields. Our attempts to go beyond what possible-N nature "allows" come with costs that are specified by possible-N. To take a simple example: If I go outside my house naked in October, I will soon feel the temperature as a limit on how long I can remain outdoors. However, I can surpass these limits by wearing clothes—a new dispositional partner that eventually enables me to engage in different behavior outside my house in October. It should be more than obvious that clothes are not randomly related to the outside temperature. As I previously argued in Section 3.3.5, clothing (in this case) acts as a "representation" of the temperature *in the relationship between* my dispositions as a (naked) human being and the relational (possible-N) dispositions of the temperature. In this way, every possible-I arrangement more or less signifies the relational limits imposed by possible-N dispositions. This goes for clothing as well as the most advanced machines (like the ASML EUV lithography machine). We do not have a possible-N order on one side and a possible-I order on the other; rather, we have a *limit arrangement* that manages the transgression of possible-N limits. Consequently, the extensive possible-I system we observe in modern Western societies roughly mirrors the possible-N limits that are transgressed. In other words, the scale and scope of our (first-order) manipulations are largely reflected in the technological, administrative and coordinating powers established to manage and maintain what possible-N requires from us if we wish to deviate from possible-N.

Thus, it is possible-N that "teaches me" to weed or to take similar measures when I want to grow potatoes. However, 'teaching' is not the same as 'determining.' What possible-N nature does and what we must do when we want to go beyond this doing are closely related but not predetermined. We may choose to weed manually or use herbicides, weed robots, mulching or other techniques. We might even decide against using any of these methods if we are less concerned about part-limiting optimization and nexus efficiency (see, e.g., Fukuoka, 1975).

From what has been explored thus far, it should be clear that our choices regarding 'what we need to do' make a difference. Significant deviations from possible-N may satisfy our isolated will but come at a higher cost, further down the line. Conversely, smaller deviations result in lower costs but also less fleedom.

Deviations within a part-limiting framework involve restructuring how the parts—originally formed and interconnected in a possible-N order—fit together. To 'pick apart what is bound' requires the effort of establishing alternative relationships—what I have referred to as limit arrangements. Only by being aware of this caveat can we begin making decisions that seek arrangements where the mix of possible-N and possible-I partners can be viewed as more or less optimal.

Without this awareness, we seem to be confined to what I have termed nexus efficiency. In nexus efficiency, we operate within a bubble of an already established limit arrangement, attempting to rearrange the presence of parts toward a higher (trivial) optimum (see Ellul, 2018, 106). There is nothing 'outside this bubble.' Thus, without a concept of possible-N, we are left with only nexus efficiency, where parts are seen as contingently related and where any notion of achieving a compromise between the (real) manifesting partners becomes meaningless.

Currently, our bubble has a global energy expenditure exceeding 620 exajoules (Energy Institute, 2024, 13), with 7 out of 9 planetary boundaries crossed, according to The Stockholm Resilience Center (2025). These figures are just numbers. We may disagree on what boundaries exist and to which degree they have been breached, but is seems hard to deny that this bubble do create something that goes by the name 'environmental problems'. The question is whether this bubble genuinely fosters freedom. Is it true that, while we face these problems, we are at least generating something we all value, namely freedom?

In the following sections, I will argue that it does not. I will argue that, for the sake of true freedom, we need to strike a balance between possible-N and possible-I.

My thesis is that when possible-I systems reach a certain level of complexity, they tend to evolve into whole-limiting systems. Whole-limiting systems, as previously mentioned, are those in which the whole exerts prevailing power over its parts. I have argued that the part-limiting cosmology developed as a counterpoint to whole-limiting cosmologies, and that the notion of freedom as 'freedom from external impediment' is an expression of this conflict.

In this chapter, I have attempted to demonstrate that 'freedom from external impediment' often devolves into a notion of freedom characterized by the 'I want x, but I don't want y' template, typically brought about by the disruption of ready-to-hand comportments. In this scenario, other partners appear as more or less contingent, enabling recompositions in alignment with my goals. This is how the part-limiting idea of "freedom" (i.e., fleedom) is realized within this framework.

The costs of these recompositions are acknowledged to some extent, but not systematically. To understand them systematically, we must also recognize our relational existence and the presence of a possible-N order. Without such an ontological perspective, it is challenging to come around to fully grasp how we inadvertently build a whole-limiting system *behind our backs*. We struggle to understand how it is possible to become *freer* (fleer) while simultaneously becoming increasingly subjected to what is perceived as *external structuring powers*.

For the affluent, this paradox is particularly elusive: 'I've never been more independent—I can do whatever I want—so why am I constantly spending all my time managing all these things and attending this endless row of meetings?' Rather than acknowledging the costs of possible-N deviations—in terms of discipline and organizing powers—it is easier to blame the state, bureaucracy or the complex laws attempting to manage the growing intricacy of limit arrangements established to facilitate our expanding fleedom. As 'part-limited citizens,' we tend to view parts as the primary components in our world, making the relational work required appear mysterious. It becomes hard to understand and explain all the 'relational labor' needed when we assume that there are 'no real relations in nature' (only objects) ('We have real work, and we have paper-pushers'). Thus, our closest explanation would often be to attribute the growth of the whole-limiting system developing unnoticed to intentional forces 'conspiring to strip away our rights and freedoms,' while the truth is that these organizing powers reflect the level of possible-I deviations.

Now, this is a theory, and it might offer little to satisfy the skeptical reader. Therefore, I will seek to support the idea of a nonchosen whole-limiting order emerging behind our backs by engaging with other similar theories.

5.6 Theories of human-made whole-limiting systems

A famous and seminal work that points in the same direction as I have argued here is the work of Joseph Tainter's *The Collapse of Complex Societies* (1988). In this book (and later work), Tainter seeks to understand the rise and fall of past civilizations. Why does it seem to be a pattern that societies, which appear to achieve 'high levels of development', suddenly disappear?

> We see cities buried in drifting sands or tangled jungle, ruin and desolation where once there were people and abundance. Who were these people, and, particularly, what happened to them? How could flourishing civilizations have existed in what are now such devastating circumstances? Did the people degrade their environment, did the climate change, or did conflict lead to collapse? Did foreign invaders bring an end to these cities? Or is there some mysterious internal dynamic to the rise and fall of these civilizations? Some of us are so fascinated by these questions that we devote our lives to studying them.
>
> (Tainter, 1988, 1)

Among the various theories attempting to explain this phenomenon, Tainter's thesis is that as societies become more complex, they become less productive. He famously refers to this as 'the principle of diminishing marginal returns.' Expanding civilizations grow because their expansion grants them access to more energy, which fuels further expansion, and so on. At some point, however, the focus shifts from expansion to maintenance, as the cost of 'holding all the parts together' increases to a level where the surplus energy gained from further expansion must be allocated to maintenance rather than continued growth. The issue arises because 'the empire' was initially held together through further expansion (increased access to energy), which is no longer occurring. The underlying reason is simply a growth in complexity:

> As societies increase in complexity, more networks are created among individuals, more hierarchical controls are created to regulate these networks, more information is processed, there is more centralization of information flow, there is increasing need to support specialists not directly involved in resource production, and the like. All of this complexity is dependent upon energy flow at a scale vastly greater than that characterizing small groups of self-sufficient foragers. The result is that as a society evolves toward greater complexity, the support costs levied on each individual will also rise, so that the population as a whole must allocate increasing portions of its energy budget to maintaining organizational institutions. This is an immutable fact of societal evolution, and is not mitigated by type of energy source.
>
> (Tainter, 1988, 91–92)

Tainter supports his theory with numerous examples that may, to some extent, provide empirical backing where mine is lacking. By this, I do not mean empirical support for the specific theory developed here, but rather to suggest that the rise in complexity *occurs without our* awareness, regardless of the explanation. In other words, diminishing marginal returns are not the result of choice, but stem from choices aimed at something else—expansion, and in my case, the energetic and relentless pursuit of fleedom.

So, how can we envision how such secular whole-limiting systems actually function? In the discourse on the philosophy of technology, there is a well-known divide between voluntarists (instrumentalists) and substantivists. Voluntarists view technology as individual tools that we freely use to achieve specific goals (for an updated version of this perspective, see Weber et al., 2013). In contrast, substantivists argue that technology is not composed of isolated tools but is part of a technological system, characterizing modern technology by this systemic nature (Ellul, 2018, 87, 90). In this context, substantivists perceive individual technologies as components of a higher-order system that has the potential to develop its own agency. This is the position that Jacques Ellul and Lewis Mumford argue regarding modern technology; they contend that (single) technologies are part of an autonomous technological *system* (Ellul, 2018, 125 ff., Mumford, 1967/1970; see also Winner, 1977). The question of whether this technological system has become autonomous is a contested idea, yet it is more relevant now than ever.

In 'Humans and Technology in the Anthropocene: 6 Rules' (2014), Peter Haff describes how the autonomy of this system (what I refer to as a secular whole-limiting system, which he terms the technosphere) come about. He explains that the "workings of modern humanity are a product of a system that operates beyond our control and imposes its own requirements on human behavior" (Haff, 2014, 127).

> The Technosphere includes the world's large-scale energy and resource extraction systems, power generation and transmission systems, communication, transportation, financial and other networks, governments and bureaucracies, cities, factories, farms, and myriad other "built" systems, as well as all the parts of these systems, including computers, windows, tractors, office memos, and humans. It also includes systems which traditionally we think of as social or human-dominated, such as religious institutions or NGOs.
>
> (Haff, 2014)

No less!

Haff's aim is to determine a set of what he calls physical rules that apply to this whole-limiting system. This system consists of parts that may themselves be subsystems. The parts are *organized*, meaning their behavior must be consistent with "the system itself" (Haff, 2014, 128). To analyze the relationship between wholes and parts, Haff argues that we should visualize the relevant parts (we are analyzing) within a three-stratum picture. If humans are the system being analyzed, we would occupy stratum II, alongside other components of similar size, such as benches, other humans and computers. Stratum I includes the components that make up the parts of stratum II, such as transistors (in computers), the heart, liver and blood (in humans), and engines and gearboxes (in cars). Stratum III encompasses larger entities of which we are a part, such as a working place, a city or a museum.

The first rule is the rule of inaccessibility: Just as I do not have direct access to my own parts (my organs), a subsystem (e.g., a university) *within the Technosphere* does not have direct access to its components, which include humans. I cannot directly access my organs, for instance; I require subsystem II parts, like surgical equipment, to do that. Similarly, the stratum III (University) level must reach me through "mediators," such as an accountant sending me a letter. Subsystem III cannot simply implement its decisions *directly*.

Rule 2 is the rule of impotence, which is essentially the inverse of rule 1. In the case of humans (stratum II) being part of stratum III, humans potentially have limited (direct) influence over this system. Haff simply argues that it would be dysfunctional if the stratum III system were overly sensitive to the whims of its parts. An exception to this rule involves (stratum II) parts that can create cascading effects, such as knocking out an electrical grid.

The third rule is called the rule of control. In cases where humans occupy stratum I (being part of stratum II and where the Technosphere is stratum III), some individuals may have more influence over stratum II than others. These individuals are referred to as leaders. However, for effective leadership, there needs to be infrastructure provided at the stratum II level, consisting of stratum I components. For instance, a captain aboard a ship requires command lines and mediating-stratum-I-components (such as a phone line to the deck or a steering wheel) to implement decisions. Parts at the stratum I level that lack this infrastructure cannot exert control over stratum II to affect stratum I. This infrastructure is established to ensure the functionality of the subsystem of stratum III, and it is the leader's responsibility to maintain this functionality. The key point is that the Technosphere (stratum III, in this case) *lacks this infrastructure because it is not a designed system.* Haff describes it as complex and leaderless, indicating that humans have no means to effectively control the Technosphere.

Rule 4 is the rule of reciprocity, which means that components within the same stratum can interact with one another. Through this rule, Haff emphasizes that the ability to interact with and potentially control elements at the same stratum—such as other people, cars, trees, TVs, etc.—does not imply that we can control or access strata above or below our own. This contrasts with the voluntarist view on technology, which suggests that we are free because we can "use the phone the way we want."

Rule 5 is the rule of performance, which states that all parts of a system must comply with the functionality of the system they belong to. For instance, going to work, paying taxes and purchasing items are all actions that contribute to the system's performance. A system as a whole cannot function without requiring a certain level of compliance. These constraints can be either soft or hard.

> A company employee experiences the hard constraints of his office, whose walls resist penetration. The door is open, but the soft constraint of fear, for example the implicit threat that he could lose his job—suffices to keep him confined for much of the day. If he is lucky, he is subject to

> the softer constraint provided by incentives, for example the prospect of higher pay or, better, the implicit incentive offered by a rewarding job – he wants to be in the office because he loves what he does.
>
> (Haff, 2014, 128)

The last rule, number 6, is the rule of provision. Stratum III, in this case, the Technosphere, is not just a system that compels its parts to perform; it also provides many goods that would not be accessible to the parts *without the system*. The Technosphere represents an environment that the parts require and partially demand. Thus, "the rules of provision and performance together create conditions conducive to positive feedback" (Haff, 2014, 134).

I am not providing more detail here, as I trust the readers ability to recognize Haff's describtions from similar ideas and from direct experience. The idea of humans being a part of a larger system has long been a part of popular culture, and anyone who has seen a science fiction film, a movie like Fight Club, or has read about structural functionalism in sociology will have some understanding of it. What is more significant about Haff's six rules is that they explain, in a more falsifiable manner, how a system that is not planned and does not have its own goals (in contrast to what is assumed in a great-chain-of-being account, for instance) *can* become whole-limiting. This is possible due to the specific dynamics between the parts and the whole. We do not need theories about malevolent individuals at the top, a slow socialist takeover, AGI takeover or invisible actors working behind the scenes to explain this. We only need a theory of the specific communication between the levels and a dynamic relationship between constraint and provision to show that a whole-limiting system can emerge without being chosen or created by intent. It can happen behind our backs.

In the argument presented in this book, I claim that our unlimited manipulations of what is fundamentally assumed to be a passive nature create a need for complex limit arrangements *due to the relational* nature of possible-N dispositions. As previously noted, this assumption of a passive nature is a significant factor in this context. When we presume that a certain type of action has a very local and/or predictable effect among the parts, while also believing we exist in a world of separated parts, we fail to recognize that our manipulations necessitate a set of activities to coordinate what has been separated at one level on the level above, which subsequently have recursive effects on the parts. Thus, the naive belief in a part-limiting freedom makes it difficult to understand why such whole-limiting systems arise.

A fourth theorist to consider is Ulrich Beck, particularly his book *Risikogesellschaft* from 1986. Thus far, we have only discussed the growth of "alternative splicing" leading to growing maintenance complexity; we have not addressed the production of environmental problems themselves and how these problems contribute to the emergence of a whole-limiting system. To some extent, I have previously struggled to show how the failure to create symmetric containers systematically creates environmental problems in each and every instance. This problem is related to our perception of plasticity in nature and what we are willing to accept as environmental problems when the benefits of asymmetric containers yield profit, wealth and economic growth. It is common to view this as a trade-off, something we consider generally acceptable, "all things considered." If this is literally *accepted* in a nonsystematic way, it is hard to demonstrate that asymmetric containers consistently create environmental problems and thus represent a systematic *cost* that originates from deviations of possible-N.

In this case, however, we need not concern ourselves with theoretical difficulties. We can simply assume that there is an increase in what is *accepted and described as* environmental problems, which is at least how Beck approaches the issue.

Beck's main thesis posits that we have shifted from industrial societies that produce goods and wealth to a risk society that produces risks. 'We are therefore no longer concerned exclusively with making nature useful, or with freeing mankind from traditional constraints, but also and essentially with problems created by the techno-economic development itself' (Beck, 1992, 19).

Beck attributes this change to the scale and magnitude of the risks. Industrial societies faced risks as well, but these were manageable. Typically, risks associated with the working environment were addressed gradually, one at a time, as industrial production generated more prosperity. Diseases related to crowded cities were mitigated through advancements in modern medicine. Risks (or what Beck refers to as hazards) are now controlled to a degree that allows us to choose whether to accept them or manage them through insurance (Beck, 1992, 21).[10]

However, the risks produced by industrial society are categorically different and must be addressed in ways that fundamentally change societies.

First, these modern risks are open-ended, unpredictable and often global in scale. Additionally, they require acknowledgment to be recognized socially (Beck, 1992, 34). They are secondhand, nonexperienced risks that depend on scientific measurements for social recognition (Beck, 1992, 71). We cannot perceive radioactive radiation, directly experience the potential threat of global warming, or fully grasp the imminent dangers posed by gene drives and genetically modified organisms. Furthermore, it is challenging to assign responsibility for these risks. While they result from human decisions, the nature of these decisions and the levels at which they occur remain unclear (cf. Haff's six rules). Previous risks or hazards were manageable, arising from chance, divine activity and intervention, or irresponsible behavior. In contrast, these new types of risks force society to confront uncertainty, responsibility and the limits of control in ways that alter its trajectory:

- As these risks increase, society becomes increasingly focused on distributing risks rather than promoting prosperity, leading to a 'risk-distributing society' (Beck, 1992, 20).
- The classical distinctions between politics and science become distorted. Science plays a crucial role in producing these risks, yet it is also our primary means of defining them, even though risks cannot be fully understood through scientific inquiry alone (Beck, 1992, 29).

- The prevailing belief that political discourse revolves around finding the *best* path to development and prosperity begins to crumble. Modernization becomes reflexive (Beck, 1992, 153), and collective agreements on societal actions are increasingly problematized and individualized.
- As nature "strikes back" in terms of risks, the clear distinction between society/culture and nature becomes blurred, a phenomenon that is increasingly evident in what we have come to call the Anthropocene.

Other elements could be mentioned, but the key point here is that Beck's understanding describes the extent to which asymmetric containers affect our society. Up until this point, I have focused on how containers (deviations from possible-N) require 'relational work' in general and how these containers may produce a whole-limiting system behind our backs. What Beck demonstrates is that the intentional or unintentional externalities originating from *asymmetric* containers change our society in further ways that cannot be neglected if we want to assess the cost and, thus, *the real efficiency of our possible-N deviations*. These costs do not surface in the equations of every company of course, but rather in the *oikos* of society as a whole, contributing to the *further* swelling of a whole-limiting order.

As noted above, Beck's account does not indicate that these costs are the result of the actual costs associated with environmental problems, but rather the perception of these problems (risks). As previously argued, it is impossible to establish the factual state of these problems, as they are not directly tangible and depend on how we view the plasticity of possible-N. Our advanced society can continue to ignore the impact of asymmetric containers (Section 3.3.5) on possible-N and renegotiate the limit arrangement (expanding the range of possible-I partners), given enough energy supply. What Beck's account illustrates are the actual societal developments that arise from the production of (perceived) risks and the associated costs.

The advice for those keen on maintaining a part-limiting understanding of limits should therefore keep up the work on shifting our common perception of environmental problems to minimize the costs of managing risks. The problem is that their audience would be precisely those who value fleedom above all else, and who will, more than anyone, be caught in the growth of a whole-limiting order that is inevitable under any fleedom scenario.

This means that even those who are *not* concerned about the environment should engage with and reflect on our core ideas about freedom and limits. Furthermore, those who are not worried about environmental issues should stop blaming individuals or institutions that advocate for respecting the limits of nature. The reality is that this situation arises not primarily from people's political views or psychological profiles ('some people are just obsessed with stopping others from pursuing their dreams in life'), but because we are gradually widening the gap between possible-N and possible-I by dismantling possible-N piece by piece, day by day, entangling ourselves more and more in second- and third-order *masked* possible-N dependencies.

Notes

1 Edward O. Wilson, with his work *Sociobiology* (Wilson, 1975), should perhaps also be mentioned.

2 'Autonomy' in this context refers to Kant's definition (literally 'self-law-giving'), which contrasts with heteronomy, where external laws or influences determine one's behavior (*Groundwork*, Section II, at Ak. 4:440). If asked, I would prefer an account *based on* the assumption of homeostasis in biological creatures. From this perspective, a better understanding of human autonomy could emerge. However, this is a complex question that would require too much attention in this book.

3 With reference to Heidegger's concept *of jemeinigkeit*, one may argue that humans are unique in their ability to relate to their own being, however. See Section 4.3.

4 In *Feminism and the Mastery of Nature* (1993), Val Plumwood argues that the classical dualism between humans and nature, as well as other dualisms, originates from what she calls 'denied dependencies.' She explains the 'othering of parts'—by constructing them as inferior and passive—which leads to dualisms, as a result of what might be termed 'our Hegelian psychology.' I am uncertain whether this view complements or conflicts with the perspective presented here. I attempt to explain 'denied dependencies' by referring to a specific historical shift toward a part-limiting view, while Plumwood seems to attribute this tendency to human nature or human psychology. This is particularly true of her account to the extent that her analysis relies on the Hegelian understanding of how self-consciousness comes into being (Hegel, 2018/1807, division B).

5 A philosopher who has tried to argue against this utilitarian outlook is Michael Sandel (2007). Not sure if his arguments are strong enough to counter clever 'thought-experiment-philosophy' if we take what was said about expansive realignment in Section 4.5 into consideration though.

6 An indication of this might be: (I)n the year 2020 (±6), the anthropogenic mass, which has recently doubled roughly every 20 years, will surpass all global living biomass' (Elhacham et al., 2020, abstract).

7 The parallel here would be the idea of falsification in science. Famously, Karl Popper argued that falsification of what is not true brings us closer to what is true. Thus, we can have progress in science. What I argue here is that this strategy is not viable in the case of progressive freedom, as the elimination of obstacles does not eliminate the relation itself. More about this in the sections to come.

8 As previously discussed, a whole-limiting cosmology became less of a compelling alternative when the concept of the "whole" had ontologically lost its normative or axiological structure, reduced instead to a collection of parts contingently "integrated by laws" (cf. determinism). Additionally, the axiological view may still pose a significant threat, particularly in the form of theocracies (prevalent in certain societies), where human freedoms are suppressed under the guise of a divinely ordained natural order.

9 One may argue that in a relational theory, 'everything is connected' so there is no fixed set of partners in any case, but this is not a problem for the point I want to get across here.

10 Beck distinguishes between hazards (or "Gefahren" in German) and risks. Hazards are typically external events, such as floods and famines, as well as traditional workplace hazards. When a hazard can be avoided through human choices, it transforms into a risk. The improvement of occupational health and safety aims to mitigate risks, not hazards. The types of risks that Beck addresses in this book are those that result from human decisions but become uncontrollable, lack clear responsibility and are often irreversible and global in scale.

References

Anjum, R. L., & Mumford, S. (2011). *Getting causes from powers*. Oxford University Press

Beck, U. (1992). *Risk society: Towards a new modernity*. Trans. M. Ritter. Sage Publishing.

Boström, N. (2005). The fable of the dragon-tyrant. *Journal of Medical Ethics*, *31*(5), 273–277. https://doi.org/10.1136/jme.2004.009035

Elhacham, E., Ben-Uri, L., Grozovski, J., Bar-On, Y. M., & Milo, R. (2020). Global human-made mass exceeds all living biomass. *Nature*, *588*(7838), 442–444. https://doi.org/10.1038/s41586-020-3010-5

Ellul, J. (2018). *The technological system*. Wipf and Stock Publishers. (Originally published 1980).

Energy Institute. (2024). *Statistical review of world energy* (73rd ed.). Retrieved from https://chrome-extension://efaidnbmnnnibpcajpcglclefindmkaj/file:///C:/Users/sno001/Downloads/Statistical%20Review%20of%20World%20Energy.pdf

Fukuoka, M. (1975). *One straw revolution*. Rodale Press.

Haff, P. K. (2014). Humans and technology in the Anthropocene: Six rules. *The Anthropocene Review*, *1*(2), 126–136. https://doi.org/10.1177/2053019614530575

Haraway, D. J. (2016). *Staying with the trouble: Making kin in the Chthulucene*. Duke University Press.

Hegel, G. W. F. (2018). *Phenomenology of spirit* (T. Pinkard, Trans.). Cambridge University Press. (Original work published 1807)

Mumford, L. (1967/1970). *The myth of the machine* (Vols. 1–2). Harcourt Brace Jovanovich.

Phillips, B. (1962). Introduction. In D. T. Suzuki (Ed.), *The essentials of Zen Buddhism, selected from the writings of Daisetz T. Suzuki*. Greenwood Press Publishers.

Plumwood V. (1993). *Feminism and the mastery of nature*. Routledge. https://doi.org/10.4324/9780203006757

Sandel, M. J. (2007). *The case against perfection: Ethics in the age of genetic engineering*. Harvard University Press.

Tainter, J. (1988). *The collapse of complex societies*. Cambridge University Press.

Weber, R., et al. (2013). The ICE-theory of technical functions. *Metascience*, *22*(1), 1–20. DOI 10.1007/s11016-012-9642-9

Wilson, E. O. (1975). *Sociobiology: The new synthesis*. Harvard University Press.

Winner, L. (1977). *Autonomous technology: Technics-out-of-control as a theme in political thought*. The MIT Press.

Stockholm Resilience Center. (2025). https://www.stockholmresilience.org

6 Relating to limits in politics, ethics and economy

6.1 In situ politics on Mars

The previous chapter concluded with the assertion that there is an inherent contradiction in our understanding of freedom. While we may believe that we can liberate ourselves by overcoming what we perceive as natural limits, we inevitably face these limits again in transformed possible-I arrangements. This transformation not only entails more work but also requires specific organizational frameworks. The limits of nature do not vanish; instead, they reappear in various less recognizable forms. This perspective challenges our confidence in the belief that humans become more human and more free as physical strains on our existence diminish. For many, it seems self-evident that greater control over our environment creates more opportunities and space for political and ethical shaping of society, that is, self-governance and a collectively chosen whole-limiting order. To illustrate this point, it may be useful to envision what a society on Mars might look like.

The idea of expanding our civilization to Mars is controversial and has many critics. There is, however, a question of whether the idea is criticized for the right reasons. It is true that achieving success in this project will be extremely difficult, but the idea is that if the optimists can prove the skeptics wrong, they may create a social system similar to ours, characterized by fewer restrictions and more freedom (e.g., Zubrin, 2022).

Mars is undeniably an inhospitable environment. Its atmosphere consists of over 95% carbon dioxide and lacks oxygen. Mars is about half the size of Earth and and has only about 38% of Earth's gravitational force. The average temperature is −60°C, ranging from approximately −130°C near the poles in winter to +20°C at the equator in summer. The absence of a global magnetic field and a breathable atmosphere exposes any life form to harmful radiation from space. Severe dust storms can alter the planet's surface in just a few days. Conditions necessary for sustaining bacteriological life to sustain (for instance) human intestinal functions are nonexistent.

If you were to arrive on Mars unprotected, you would immediately experience the detrimental limits of the environment. There are virtually no friendly

DOI: 10.4324/9781003307372-7

elements to support human life initially. Consequently, the number of dispositional partners that have to be put in place to sustain human life would be monumental.

The primary concerns raised by critics of Mars colonization are largely technological: How can we sustain plant life? How do we terraform the Martian surface? How can we access water? How can microbial life survive in high radiation? How do we mine and produce metals? Life on Mars will predominantly involve maintaining the structure of possible-I dispositions, which will need to be immensely massive, and those who venture there will be acutely reminded of their relational nature in a way that we only partially experience on Earth. But let's assume these technical challenges are resolved: What might the dominant social conditions in a Martian community be?[1]

The scale of the technosphere and the systems required to coordinate it will undoubtedly be vast. The state's role would not be to secure freedom but to ensure the established possible-I order. Scientific predictions would have to be highly accurate, with minimal margin for error across a wide range of scenarios. A significant portion of Martian activities would need to be instrumental, with practically everything serving a specific function, rendering life's givenness completely humanized. Responsibilities would be strict, and almost any action could potentially harm others. The degree of specialization, division of labor and hence the need for coordination will need to be extensive. Irrational or unpredictable behavior would need to be minimized and controlled. Reproduction and upbringing would require strict and unified protocols. Everyone would need to perform their tasks as designated by the overarching system.

This is, of course, speculative and may not accurately reflect all aspects of life on Mars. Readers will surely find better and more precise descriptions of the nontechnical issues associated with human life on exoplanets, but I think it is safe to assume that politics on Mars will not be quite the same as a libertarian would call ideal.

6.2 Politics on earth

What I want this trip to Mars to illustrate is the connection between politics and our manipulations of nature. Those (societies) that decide to build large possible-I systems, such as the Mars colonization, will eventually become more dependent on the function of the system than those who go for more moderate limit-arrangements. Settling on Mars requires the construction of large possible-I systems, where collaboration, structures and institutions relating these limit-arrangements are a matter of life and death. In such a scenario, collaboration must be integrated; things need to happen more or less automatically without negotiation. A minimal libertarian state cannot exist on Mars. If a creative individual, exercising her free speech, suggests do this rather than that, there will be a direct experience of how this must lead to (noncontingent) changes in a range of different parts of the system. What will the maneuvering space for creative destruction look like on Mars if most changes are a matter of life and death? One may ask: What is the premise of the liberal state if not the relative friendliness and plasticity of possible-N on earth? A friendliness that we today seem to exploit rather than appreciate.

As humans continue to build a civilization on Earth—through economic growth—that depends on limit-arrangements which vastly deviate from possible-N, while simultaneously making the Earth less inhabitable through environmental degradation, these systems might also become increasingly less negotiable. In other words, they might, like in the Mars scenario, become less friendly to the idea of a liberal state.

Just imagine what it would take if the New York City Council decided to dig up and replace the entire sewer system. How many authorities would need to be consulted? How many decisions would need to be made? How many meetings would need to be held? How many different types of manual workers would be involved in a discontinuous line of operations? If such a political and bureaucratic system is not in place, this will not be possible. Should we put the blame on nonliberal politics?

From a liberal standpoint, the assumption is that the more we control our limits through possible-I systems, the wider the space left for 'free political decision-making,' personal freedom and ethical deliberation. However, this seems mistaken. As our deviations from possible-N grow, the importance of coordinating forces instantiated by bureaucracy and "politics" increases. This is not 'free politics,' but rather reactive politics. If everything requires more coordination and integration to function, there is less room for critical thought and fresh ideas. In a whole-limiting order no one controls (e.g., the technosphere) the best ideas will, in all cases, be those that solve nexus efficiency problems. If a governmental body says that 'Norway will have a deficit of 26 TWh by 2040,' there is no debate on whether this is taking us in the right direction. There might be considerable debate about this issue in the public sphere, but the *political task* is to find a way to increase Norway's energy production accordingly.

The Mars scenario illustrates a whole-limiting system where individuals become entirely dependent on the functioning of the system itself, where it is in one's self-interest to let the system 'do its work.' By analogy, our bodies allow no 'free politics' either. The heart does not choose its own way of life, nor should it, because if the liver did the same, it would jeopardize its own existence. Under such conditions, the original liberal idea—that I can do whatever I want as long as it does not limit others—becomes meaningless, since there is rarely anything I can do that does not affect others in such systems. Thus, the only way I can comply with this principle is to adhere to the *functionality* of the system.

At the same time, as the technosphere expands through a steady production of unintended consequences (asymmetric containers) and a growing range of goals made available by increased means, politics also becomes more "important." What was previously just 'happening by nature' now becomes a matter of political and ethical choice. Where fishermen once attributed the absence of fish stocks in the sea to God or destiny, they now hold scientists and politicians responsible. Nature becomes politicized. This process is well-documented in Anthropocene literature. What I will add or emphasize here is the shift in responsibility. When nature becomes *normature* in this way, one implication is that more of the processes occurring in nature also become normed. The temperature in the air, hurricanes in the Philippines, rising sea levels and the rapid disappearance of species are events and processes that trace back to human "decisions." This can be seen as a rebirth of axiological teleology—the idea that there is a goal in nature permeated with moral value. The difference is that these norms are not produced by God or 'nature itself,' but by humans. We are placed in a position where we must decide 'what kind of nature we want.' We become 'The God species,' as Mark Lynas calls it. When nature becomes normature, what kind of nature we have becomes a human choice.

Normature and the rebirth of axiological teleology ironically result from our part-limiting freedom, which takes other nonhuman parts to be contingent obstructions to our freedom. The question is: What are the chances that our *political system* can choose correctly when asked, 'What kind of nature do we want to have?' What does a democratic nature look like? If, let's say, we choose a particular (dynamic) state of nature that keeps the integrity of the main ecosystems on earth intact, to what degree is it possible to decelerate human expansion to reach this goal?

6.3 Agreeing on a chosen whole-limiting order

In this book, I want to contribute to the ongoing debate on liberty and the environment. I hope the previous chapters have helped explain why we should not approach this debate assuming that there is an inherent conflict between environmental concerns and freedom. This assumption, I argue, comes from a part-limiting liberalist conception of freedom, which sees 'the environment' as external forces that threaten or restrict our freedom. I have also tried to show that those who adopt this part-limiting liberalist conception of freedom are trapped in a paradox: the more we pursue this freedom (fleedom), the less free we—and all of us—will become. I believe this should serve as our starting point for a conversation on politics and the environment, which is not currently the case.

What we see, rather, is a consensus that there is a conflict between freedom translated into economic needs and environmental concerns. It is often framed as "growth versus protection" and "employment versus the environment." Restrictions on mobility, such as air travel, are viewed as threats to our freedom. As a result, environmentalists are perceived as "party poopers" intent on 'undermining the country.' The academic discourse on environmental policy also seems to, quite narrowly, take the part-limiting idea of freedom as a starting-point (Hannis, 2017, 6).

If we delve deeper into this debate on freedom and the environment, it is typically framed as a contest between liberals (neutralists) and communitarians/republicans (perfectionists).[2] (Note that a relational understanding of freedom is absent.) The perfectionist is given the role of finding common ground for an account where 'the environment and our human freedom' is less in conflict. But 'less in conflict' means, on a part-limiting view, to limit the freedom of some groups or individuals. Thus, environmentalism becomes associated with a stance that requires arguments that, from a neutralist perspective, impose a *particular view* of what constitutes a good life on everyone else (i.e., what is usually associated with 'positive freedom') by default. This is a starting point where care for the environment can never win.

The underlying issue is that perfectionists are not relationalists, but rather fundamentally liberal as well, or so I claim. This places the onus on the perfectionist to demonstrate that environmental concerns do not entail a vision of a good life that eventually is imposed on others' freedom to choose. However, this is impossible to establish within a part-limiting framework, as this view holds that 'the environment' consists of contingent parts (no distinction between possible-N and possible-I) in the first place. There is nothing within the environment that can legitimately restrict my freedom, except for those limits that are clearly insurmountable. Trespassing limits that are surmountable (relational limits as defined in this book) is the type of limit that belongs to the realm of individual choice in the outset.

Why is this so? Within the context of a part-limiting view, whole-limiting orders are not found or sought, (e.g. the Greek-medieval whole-limiting order) but rather *chosen*. The way human history has worked out, this chosen "whole-limiting" order has been called the democratic state.[3] Since it is supposed to be chosen, we often engage in fierce debates over what this order of the state is and should be. The options range from anarchism to certain forms of fascism. Thus, within a part-limiting view, the basic contender is not perfectionism, but a (naturalized) whole-limiting order! When we agree that a (naturalized) whole-limiting order is our common contender, classical liberalism serves as the default position. Politically, the classical liberal view of the *social contract* (between individuals that agree to create a state) posits that the state (the chosen whole-limiting order) should only regulate tensions and conflicts among those who have willingly chosen to be part of the order. This is the essence of the contract: the individual exists not for the benefit of the whole, but the whole exists for the benefit of the individual. The state is 'at service.' But what is it in service of? In Thomas Hobbes's view, individuals are equally "free to do whatever they want" (i.e., possess natural rights/liberty), which leads to the infamous "war of all against all." In this scenario, there is no (whole-limiting) natural order there *in the first place*.[4] Choosing the artificial order of the state becomes the most rational option. Consequently, the state's obligation is to protect the natural rights of individuals, which would otherwise be threatened by other individuals. The only *natural* unit here, if any at all, is the part: the individual.

As previously argued in Chapter 2, the process of entering a chosen whole-limiting order that secured the equal rights of individuals unfolded over a long period and involved replacing a natural hierarchy (e.g., the privileges of the aristocracy) with a type of intervention in nature suggested earlier by Francis Bacon. (See e.g. the excerpt from Ingersoll's speech in the Preface.)

For there to be equal liberties for all, it appears—at least historically—that there is no way to achieve such liberties without the Baconian subjugation of nature. The instrumentalization of nature seems to go hand in hand with the establishment of a chosen, whole-limiting order. This appears to be true, at least, when a society is halfway to accepting that it exists in a part-limiting world, where the struggle for freedom is defined as a conflict against other parts, both human and nonhuman.

What this means is that the perfectionist, no matter how close it comes to acknowledging our relational nature or our 'deeper connections with nature' cannot appeal to any nonchosen order beyond the individual. It is only the individual that has 'natural rights.' Hence, any attempt to place humans within a larger whole or a relational web of dependencies will be interpreted as an attempt to subject everyone else to the same idea of special interests.

So what does it truly mean for the state to protect our natural birthrights? As previously asserted in Chapter 2, political liberalists have famously argued for a distinction between formal rights and the individual right to choose how to lead a good life (neutralist stance).[5] The state should not hold any opinion on what it means to be human or how to achieve a good life (the perfectionist idea).[6] An idea of "what humans really are" can easily become a tool for groups seeking to advance or consolidate their special interests, undermining the agreement that a whole-limiting order should be chosen. If the state favors one group above the other, we are heading back to the situation that Hobbes rebelled against: those with contingent power interests that have convinced everyone else to find their place within the natural order.

But is this really an argument against the perfectionist or against the nonchosen whole-limiting orders? Does the perfectionist advocate for a specific concept of human nature, as suggested by the neutralist: 'There is a human nature, and the state should protect this nature.' I would argue that this is mistaken. The perfectionist is also a liberalist.

Instead, the disagreement revolves around a realistic description of what it means to give individuals the opportunity to exercise their rights and lead a good life. From this perspective, one could even argue that Marx was "a liberal," or at least that he uses liberalist arguments against the capitalist society. For instance, Marx typically argued that proletarians are not provided with equal opportunities to exhibit their rights and pursue a good life; their autonomy is compromised. When the state fails to recognize the power dynamics in a capitalist society, it effectively favors a particular group and their concept of the good life, rendering the state unjust—not inherently unjust, but unjust in its violation of the common idea of autonomy.

Similarly, the debate between liberalism and communitarianism does not question the *fundamental* liberal idea. A communitarian emphasizes that our notions of the good life and our life purposes are often not individually constituted but are instead sustained and made viable by the collective aspects of our lives. Exercising our conception of a good life involves elements such as language, rituals, celebrations, cultural habits, consumption of goods and knowledge, all of which are collective. Particularly for individuals belonging to a minority within a state, these realizations of the good life require protection—not just in terms of permission, but also through active nurturing (Taylor, 1992, 43). Does this contradict the basic liberal idea?

Even though the communitarian (perfectionist) emphasizes that humans are relational and social beings who depend on others, the focus remains on what the state should do for the *independence* of the individual. From a communitarian perspective, the state should do *more* to enable *the individual* to maintain their *chosen* 'social individuality.' The key difference lies in how much is required for an individual to achieve independence. We are not born autonomous. In both cases, there are *external constraints* to overcome, but since the communitarian views the individual as part of a series *of relationships that are not optional from the* outset, the state's role will differ. Women and men growing up in a patriarchal society have not chosen their relational dependencies. Similarly, transgender individuals in a transphobic culture are not free to express and live by their personal ideas about how to lead a good life.

This may seem a bit confusing, but a way to make the picture clearer could be to distinguish between liberalism1 and liberalism2. This move intends to question the narrative that perfectionism represents a position that is able to challenge the neutralist position on the problems that environmental concerns pose for our idea of freedom. There is too much overlap between the neutralist and perfectionist perspectives for a genuine challenge to occur. Both positions align with what I call liberalism1. Liberalism1 asserts that the state's goal is to serve individuals and their independence, rather than the other way around. This implies that the state must have valid reasons to interfere with what individuals consider worthy goals. Our dependencies are essentially contingent, and ideally, all our dependencies should be chosen. This is a point of agreement between both positions.

In contrast, liberalism2 claims that we are *born with* autonomous agency, and that the state's attempt to equalize opportunities for everyone represents a fundamental misunderstanding. This often leads to unjustified interference, privileging certain conceptions of the good life over others, they say (see Hannis, 2017, 7). Proponents of liberalism2 worry that a state seeking to 'eliminate all differences' will *act on behalf* of individuals in a way that undermines what they perceive as our true autonomy, effectively becoming a womb that substitutes for our inherent autonomy and alienates individuals from their genuine struggle toward freedom. Consequently, the chosen whole-limiting order must be as restricted and minimal as possible, intervening only when an individual's freedom clearly threatens the freedom of others (the harm principle). Conversely, the perfectionist view 'the active state' as an instrument for the possibility of becoming truly independent and autonomous.

From this shared perspective (liberalism1), the world appears to be one of contingent limits. Transgressing these limits thus emerges as something inherently good. This is not merely a matter of shared values; it reflects our basic understanding of the state. The state is in the business of protecting humans as elevated parts in a world of external, unrelated (allohuman) parts that, in turn, can be organized, displayed and apportioned according to our wants and desires. If the perfectionist did not share this view, their arguments would appear quite different.

From this common ground, it seems clear that when environmental concerns are translated into environmental politics, what limits us is external and contingent from the outset. Hence, it becomes challenging to argue that the natural environment and care for it are anything more than a value-based priority that *some* people hold while others do not. The threat of a nonchosen (naturalized) whole-limiting order looms in the background and it seems unjust for those who value the natural environment to decide that everyone should share the same perspective. People live vastly different lives and thrive in various environments. Some even lead lives that closely resemble what we would otherwise consider a completely deteriorated natural environment. From a liberalist standpoint, it is difficult to determine at what point the environment is deemed unsatisfactory.

This leads to the conclusion that the neutralist and the perfectionist are not on equal footing in arguing for environmental concerns. The neutralist (liberalism2) can argue that in a society where no one cares about the natural environment, it would clearly be unfair for the state to prioritize these concerns. In other words, the agreement on liberalism1 (originating from the part-limiting cosmology) compels the perfectionist to *defend* their position against liberalism2. Not vice versa. Goals that entail protection against environmental deterioration become an unjustifiable type of limitation to other people's freedom. This would not (necessarily) be the case on a relational account. The reason why the perfections bear the burden of defending politics that restrict human freedom *is the shared understanding that relations and dependencies are contingent from the outset*. This shared understanding has its origin in the part-limited cosmology.

A possible response often begins with certain quasi-relational arguments, suggesting that humans are tied to the natural environment through capability requirements that must be met. However, these theories lack consideration of the conditions in the allohuman world. There is no *ontological theory* connecting capabilities and what these capabilities point toward. Any relation to an intact ecosystem or access to food and experiences can be replaced by possible-I partners. If you want to walk in the woods, what is wrong with plastic trees? (Krieger, 1973) If the tap water is polluted, you can drink bottled water. The question of possible-I partners is merely a matter of energy supply. If all our needs and capability requirements can be substituted by possible-I systems, it will also benefit the economy. Additionally, we will see what people truly prioritize when these "relational goods" are commodified. We no longer need to consult philosophers; we can simply ask consumers.

6.3.1 Limits in nature—in politics

Since communitarian and perfectionist arguments are grounded in liberalism1, the only argument that I can see as a rejoinder to liberalism2 is the arguments put forward in Chapter 5. This would typically be an immanent critique aimed at demonstrating that the position is internally self-defeating. The limits we transgress are not eliminated but rather transformed. Thus, limits is *not* only a matter of energy; we create these possible-I systems based on an already existing *relational* possible-N system. As a result, the concept of a *chosen* whole limiting order is undermined by the continuous growth of a nonchosen one.

However, this is not my primary argument at this point. My main contention here is that the description of human nature is fundamentally incorrect, and we should not base our politics on a false preconception of what it means to be human.

Within the realm of part-limiting cosmologies, we will understandably never fully address this question: What are humans really? What is our nature? The reason is of course that this would simply be seen as purely normative discussions, which would amount to the naturalistic fallacy; of moving from 'is' to 'ought' (Lie, 2016, Chapters 1.3 and 1.4). Seen from a part-limiting perspective, to discuss human nature as if it could be definitively pinned down would be awfully naïve. Of course, differing views on human nature exist within the part-limiting cosmology: Hobbes argues that we are selfish, scheming and brutal creatures unless constrained, while Rousseau believes we are sociable, kind and naturally inclined toward compassion in our original state, albeit corrupted by society. However, these are viewed as *normative* perspectives, thus giving neutralism the upper hand.

Based on the discussions in Section 6.2, I believe it is possible to explore human nature beyond normative definitions, however. The claim that humans have a relational nature is not a normative claim; rather, it is an ontological assertion that opposes the *already established* ontological claims of part-limiting cosmology. This claim is not derived from or grounded in science; it is a philosophical claim—specifically, an ontological one based on the theory of dispositions developed into what I have called relational realism.

Much has been said about this theory in previous chapters, and I will not attempt to outline what a relational realist perspective on humans could mean for politics. Such an outline would also be highly questionable since the implications of this theory may be as broad as those stemming from the ontological claim that a human being is a self-constituted, independent part among other separate parts.

Nonetheless, some broad assumptions can be articulated. For instance, we cannot base our political system and the social contract on the assumption that freedom means to become as independent as possible. Any choice comes with a 'package deal.' Limits are not external; they are relational. An individual's capabilities and goal seeking are both enabled and constrained by their tangled partners. If we can never escape our dependencies, it makes little sense to claim that environmental concerns threaten our freedom, since the freedom in question (fleedom) never truly existed. The choices we face are not between limitation and freedom, but rather about which 'package of dependencies' we are willing to accept. The state should not be obligated to support a system underscoring the template of 'I want x, but I don't want y,' at least not by default. Instead, it should endorse the template of 'staying with the trouble.'

To challenge the situation where the onus is on the perfectionist to defend politics that restrict human freedom (part-limited freedom), we do not need to propose a fully developed theory of the relational nature of humanity. It is enough to say that 'we are relational'. This shifts the burden of proof to the neutralist, who must then justify why certain expansive actions do not infringe upon our 'relational freedom'.

The fact that our choices cannot be isolated also makes it difficult to pinpoint where and when we violate the harm principle. This suggests that many of our choices regarding transgressing natural limits are more political than we typically perceive, as many of our transgressions typically take part in producing a whole-limiting order behind our backs. In this context, we can either deny the relational ontology of humans (clinging anxiously to liberalism2) or conclude that if we desire a society with less interference from politicians and bureaucracy, we as a society should deviate as little as possible from possible-N: We should simplify and reach for optimal limit-arrangements (cf. The Great Simplification: https://www.thegreatsimplification.com).

Regarding the topic discussed in Section 1.5, we may also recognize that the definition of sustainability will never yield a satisfying conclusion for those who wish to limit the destruction of our natural environment. The crucial role of 'needs' in the definition is dictated by liberalism, where liberalism2 has the upper hand in defining the individual as the sole legitimate source capable of answering this question. This individual is one that views anything that limits her needs (preferences) as a threat to her freedom and has, as argued throughout the book, a fundamentally misconceived understanding of limits.

Keeping this debate at the forefront of environmental politics—by attempting to distinguish between needs and preferences (O'Neill, 2010), defining needs as capabilities (Rauschmaier et al., 2010) or arguing that some liberties are more valuable than others (Stephens, 2001)—will only benefit liberalism2. There are no limits in nature (other than energy supply) if the relationships between our liberties, needs and capabilities concerning the natural world are contingent. In this regard, liberalism is correct. Again, what is wrong with plastic trees? If we do not articulate a different ontological understanding of what it means to be human—one that attempts to clarify our noncontingent relatedness to our dispositional partners—our limits will only encompass those that have not yet been scientifically "debunked" and rendered technically and economically feasible.

Generally, it is hard to see how politics can be able to control and steer technological development. What we experience as a limit does not become apparent until it is conspicuous, and it does not become genuinely conspicuous until it is possible to transgress (see Section 4.5). Usually, all these steps challenging limits in nature happen before the public can have a say about whether the technology should be a part of their (chosen) whole-limiting order. Who decided that social media should become 'the infrastructure of our lives?'

Moreover, from a liberalist1 perspective, the burden of proof falls on those who wish to restrict the rights of individuals or organizations seeking to exploit the transgression. For example, as soon as we find a way to resurrect the mammoth, the limit becomes contingent, and who can justify political measures to stop someone who wants to resurrect it? The limit is not real anymore.[7] It is contingent, and to politically restrict someone from pursuing this possibility would unjustifiably hinder their pursuit of a good life.

6.4 Endnotes on politics

Many environmentalists put their trust in the level of politics to solve the environmental crisis (e.g., Scavenius, 2018; Vogel, 2015). Environmental problems are problems that cannot be solved on an individual (or national) level, one would then argue. I would argue, however, that it is naïve to simply assert this as an obvious claim without also pointing to the specific challenges that (part-limiting) politics is up against when it comes to the relation between the human and the natural world.

First of all, there is a nonchosen whole-limiting order originating from our deviations from possible-N, growing in parallel with the chosen one, which to a large extent outcompetes and displaces the space for real political choices. As the leeway for further human expansion becomes narrower due to the vast deviations from possible-N, further expansion will put politics in the service to coordinate, restrict and control our individual activities to maintain the system functioning of society and not in the service of the natural environment or the individual (cf. Haff's sixth rule of provision). What politics on Mars would look like is supposed to illustrate this point.

The second problem is that the very idea of politics is built on an assumption about limits that makes it impossible for politics to limit further environmental destruction. When the purported conflict becomes acute, 'environmental concerns' will always be considered illiberal, threatening to overthrow the very idea of a chosen whole-limiting order. What we could hope for is that political decisions coordinate our possible-I deviations in a way that is grounded in limit-arrangements which are *genuinely* efficient and do not lead to the perpetuated growth of asymmetric containers and the nonchosen whole-limiting order. But for this to happen, we need to acknowledge that limits are relational and that this relationality is tied to a possible-N order. This is not a small matter. Science would need to acknowledge the existence of such an order, and the upper hand of liberalism must be challenged. The part-limiting cosmology would need to be replaced by a relation-limiting cosmology.

6.5 Ethics

As noted in Section 1.3, our assumptions about what we can do are often regulated by what we ought to do. "Ethics" and "values" are integral to our lives, de facto limiting our behaviors. For instance, I may feel inclined to harm my neighbor because I dislike him, or lie to my friend about missing her birthday. However, I refrain from these actions for ethical reasons. The critical question is whether there is a connection between these ethical limits and the limits found in nature. At first glance, we might dismiss this question by invoking the is/ought distinction: Whatever limits we may find in nature, they have no bearing on possible-V (V for values), that is, what we should consider as binding in the ethical realm!

Building on what we have discussed earlier in this book (Section 2.4), we can recall the reasons for denying that limits in nature should influence the ethical constraints on our behavior.

The scientific revolution, along with its philosophical foundations, was a reaction against the axiological cosmology that posited values as inherent in nature. In this view, nature and its parts strive for the good (see Section 2.3). Modern ethics, shaped by a part-limiting cosmology, focuses on the individual actions of these parts. The good must be identified within the *parts, and only certain parts can qualify* as possessing ethical value (see Section 2.5.1). Ethical capacity are located within the parts, and only some of these parts are capable of acting ethically.

As previously noted, this relocation of ethical value traces back to the passivist and mechanist views of nature (Section 2.4). The 'agency' of natural parts does not stem from an inner drive; rather, these parts are acted upon by extrinsic forces. What occurs in nature is determined by antecedent causes, which are either governed by (contingent) laws or, if not laws, are simply contingent (cf. modal dualism). Therefore, for any part to qualify as possessing ethical value, it must have intrinsic properties that can modify the influence of these extrinsic forces. Only agents with the appropriate intrinsic properties can act freely and ethically. Other parts cannot be held responsible because they could not have acted differently. Since we cannot assume or justify the belief that 'nature and its parts' strive toward the good, we can only attribute 'the good' to those parts capable of choosing between good and bad actions. This ultimately means that the ethical realm aligns with the human realm. Thus, to argue that limits *in* nature should impact ethics would be a categorical misunderstanding; limits in nature are simply not (directly) relevant to our ethical reasoning. This, I have argued, suggests that the modern concept of intrinsic value, seated in the parts, has its origin in the passivist ontology of the scientific revolution. To the extent that one accepts this ontology—whether implicitly or explicitly—one might view those who ascribe ethical value to allohuman nature as committing a fallacy of reasoning, specifically the naturalistic fallacy.

However, environmental ethics has famously challenged this anthropocentric assumption, which Hans Jonas terms "island ethics": the notion that only those parts capable of enacting responsibility are worthy of ethical consideration.[8]

The strategy to expand the island of entities to which normative value can be ascribed makes sense within a part-limiting cosmology, but only if we can navigate the issues that gave rise to the concept of intrinsic value in the first place.[9]

My claim is that environmental ethics, particularly the attempts to expand the range of entities that should be included, have not sufficiently addressed the ontological ties to passivism (Lie, 2016, chapters 1.3, 1.4, 5.7, and 2021). It is not properly understood that if you want to challenge the island ethics you are not only in the business of arguing that the realm of ethics should extend to parts endowed with other properties than what we supposedly find to be exclusively human, but also in the business of challenging the general scientific *description* of nature (ontology). More specifically, the question arises: by what *authority* can an environmental *ethicist* argue that the behavior of whales, for example, is not merely the result of an antecedent causal chain of events (Lie, 2016, 34)? The only way for the ethicist to challenge the passivist ontology of science is to engage with ontological questions. To be fair, some have pursued this path, but perhaps not in the way that the account given here would ascribe (see, e.g., Callicott, 1985; Rolston, 2003, 145).

A more recognized, yet notorious, problem for environmental ethicists seeking to expand the realm of ethical worth is explicitly related to part-limiting cosmology. This cosmology frames ethical value as a question of where value resides. Does it reside in the part that values (subjectivism, emotivism, etc.) or in the valued part (objectivism) (Palmer, 2003, 17)? Do we value something because it is inherently valuable, or because we project (ethical) value onto it (cf. Plato: Euthyphro Dilemma)? While this debate is challenging enough within the realm of human-centered ethics, it becomes even more complex when attempting to extend beyond the human realm.

What is often misunderstood here, I think, is that this problem arises from the part-limiting cosmology, and that the famous subject/object divide, as well as the human/nature divide, are instantiations of this cosmology—not the other way around. Due to this misunderstanding, we find ourselves constrained when trying to extend ethical value. To extend ethical value means to extend *intrinsic* value, which implies an absolute value. Something either has (ethical) value or does not. If something possesses intrinsic value, its value must be fully respected; it holds value in itself. If it does not, it has only instrumental value, which *from a passivist* perspective translates into treating these parts as if they 'act' solely due to antecedent and determined and/or contingent causes. In that case, the only limit to our conduct would be what is 'causally possible.'

Thus, extending moral worth means that the selected parts (such as higher mammals) are *not considered part* of a passive nature but instead become "untouchable," like humans. Acting ethically under these terms requires that we leave the selected parts of nature completely untouched. It is an ethic that forces us to view ourselves (and the selected extensions of parts) as separate from nature, thereby instantiating a dualism that most environmental philosophers seek to overcome through their criticism of island ethics in the first place (e.g., Vogel, 2015, 21). If the extensionist does not confront the underlying ontological assumptions of passivism, which require intrinsic value as a marker of ethical value, they will ultimately reach a point of unresolved exhaustion. Conversely, a thorough critique of passivist ontology will lead our thoughts in a direction where the struggle to extend *intrinsic* value becomes self-defeating.

As previously argued, axiological cosmologies (at least in their Aristotelian version) and the idea of intrinsic value share a common denominator. If something possesses goal-directed features, we may ascribe ethical value to it. However, passivism puts a preassure on humans (philosophers...) to narrow the ontological range of goal-directed behavior. The issue with this narrower range is not only the risk of 'anthropocentric ethics' but also that it is rooted in a passivist ontology, which undermines the possibility of any connection between limits in nature and ethics.

As the reader engages with the arguments presented, some may challenge my criticism of intrinsic value by referencing the debate over its true definition. I am particularly thinking about John O'Neill's seminal paper, "Varieties of Intrinsic Value" (1992). Following the arguments in this paper, it could be argued that I conflate the different meanings here.[10] My response to this objection is as follows.

I base my argument on a historical exploration rather than an *'empirical' analysis* of how the concept is actually used. I aim to understand why intrinsic value holds its place in modern ethics, especially in environmental ethics. I attribute this significance to the idea of intrinsicness. First, I examine how it became momentous in ontology. The broader context involves part-limiting cosmology, while the narrower aspect focuses on how passivism has shaped our understanding of intrinsic versus extrinsic properties.

For instance, stating that a tree is ethically valuable *implies* that the tree possesses some intrinsic properties that carry value (e.g., rationality or consciousness). Since science cannot confirm that a tree has such intrinsic properties, it is easy to demonstrate that this claim is a fallacious projection. The value does not reside in the tree but in the valuer. The argument is that the ontological concept of intrinsicness spills over into ethics, leading us to seek a way to verify that the property we discuss is indeed intrinsic (objective): that is, independent of everything else (nonrelational). If we could establish this, it would imply a *noninstrumental* value. When O'Neill (1992, 121, 122) argues that emotivists could attribute noninstrumental value without suggesting it has objective value, he is already operating within the intrinsicness framework (subjective/objective values).

The central point here is not the different meanings of intrinsic value, but the fact *that* we are trying to identify in which part value resides. This unites all three meanings as expressions of the same "cosmology." When I conflate these different meanings, I express the same basic assumption: that values are fundamentally not relational. It is only when we recognize how intrinsic value and relational value differ that we can understand that the various meanings are *not* in conflict. If you insist on the intrinsic value account, you will find O'Neill's analytical distinctions valuable, but not if you view relational value as an alternative to intrinsic value.

Before we move on, it is important to grasp the scope of what is here meant by "any connection between ethical limits and limits in nature." This could mean that 'natural law' (*phusis*) should dictate 'human law' (*nomos*). This relates to what I have previously referred to as (unsound) naturalization, with one of the oldest examples being Thucydides' Melian dialogue. The dialogue describes a situation where the Athenians threaten the Melian population to either subjugate or face annihilation. Rather than referring to the current normative standards of such negotiations (*nomos*), the Athenians invoke 'natural law' (*phusis*), where 'natural law' in this case means 'might makes right.' They (the powerful) use what they *interpret* as "natural" to achieve their goals among humans.

This is not what I mean by connecting limits in nature with ethics. The scope of our investigation is whether the proposed idea of limits in nature should influence our ethical limits toward allohuman beings at all (Lie, 2016, 1.4.1.c). If it does, this will of course affect our perceptions and actions, but not in the sense of "nature-derived" norms for how we should treat each other. We are not seeking an account where norms can be *derived* from descriptions.

We need to remember that in a part-limiting cosmology, it is very difficult to pin down whether something follows from anything at all. In logic, we learn that the conclusion cannot entail more than the premises. To assert that "ought does not follow from is" is merely a truism. If everything is composed of discrete entities, the proposed basic relation would be causal, meaning that the noncontingent relation between these entities would essentially be a leap of thought, leading to the problem of induction. To take the naturalistic fallacy seriously, we must accept the part-limiting view while recognizing that, within this framework, nothing necessarily follows from anything in the first place. This principle is not ontologically neutral. It is therefore not an illegitimate move to assume that a different ontology could potentially reshape our understanding of the relationship between facts and norms. However, this does not imply that a different ontology will create a situation where norms necessarily follow from facts. The concept of necessity itself is part of the part-limiting ontology, as the only alternative available would be contingency. What we can assert with certainty is that a relational ontology cannot provide an account where values are intrinsic; they must be relational. After all, if everything else is a mutual manifestation, it would be strange to accept that value must be a manifestation of one discrete and nonrelated part (i.e., a manifestation of either the subject or the object).

Thus far, the focus has been on exploring an alternative account of limits in nature. If we accept this account, we may also recognize the necessity of reassessing how we act within these limits. Expanding beyond the limits in nature implies that we seek the right compromise, consistent with what has been discussed regarding the details of such a compromise (Sections 5.4 and 5.5). When we consider ethical limits, we are *additionally* questioning whether we should engage in such a process at all. We may seek a compromise when we want to install wind turbines in a certain area, *but should we even put up the* wind turbines at all?

From what we have discussed thus far in this chapter, it should be clear to the reader that we cannot (directly) seek help from traditional (part-limiting) ethics or from the whole-limiting cosmologies of the past. If we are serious about bridging the gap between limits in nature and ethics, we must do so without relying on arguments that depend on intrinsic value or cosmological axiology. This is a challenging task, and the aim of this section is not to find a patent solution to this problem, but rather to explore and consider *possible outlooks and positions* based on what are commonly referred to as relational values.

6.5.1 Relational values

A relational ethics relocates normativity and denies the idea that normativity is to be sought primarily *in the part*. Normativity rather emerges from mutual manifestations, so to speak. There have been clear traces of a more full-blown account of relational values in the past. In 'Intrinsic Value, Quantum Theory and Environmental Ethics' (Callicott, 1985), Baird Callicott argues for a relational view of the world in general and that quantum physics collapses the subject/object distinction, which for him (being a Humean) means that attributing value through our feelings (e.g., sympathy) does not mean that value resides in the subject and that the object has no value beyond the attribution. Holmes Rolston (2003) argues that value emerges relationally with the appearance of the subject-generator (144). Nature is a value-producing system, making it possible for us to value it. Value emerges between the valuer and the valued in the first place. A more explicit and perhaps the most seminal account can be found throughout the work of Arne Næss (although based on a more suggestive relational ontology than the one developed in this book. See Brennan & Lo, 2021, Lie & Wickson, 2011. He argues that subjects are not placed in opposition to objects, but within what he calls a relational field (see also Section 6.5.3.3.2).

Interestingly, the relational view on ethics has also had a growing traction over the last few years, especially within the environmental discourse (Chan et al., 2018, A1).[11] These accounts differ from mine primarily in the ontological dimension: I emphasize that a proper relation ethic must explicitly challenge the ontological foundation of property-based ethics and find a more solid base in a different ontology.[12] Next, I offer an account of what seems to be the current 'relational values account' and then I move on to explore what could possibly be a proper understanding of the relationship between relational realism and ethics.

What is today called 'relational values' looks something like this: First of all, proponents of this interpretation of relational values are in most cases not from the traditional population of environmental ethicists (Chan et al., 2018, figure 2). Some aspects of the environmental ethics narrative have changed. A typical example would be that a seminal reference within this new school is T. C. Brown (1984), while among environmental philosophers it would rather be the names mentioned above (see also Chan et al., 2018).[13] Contributors to the relational account in ethics are more interdisciplinary and more focused on ethics applied in real-world situations and less on 'theoretical hairsplitting' in the field of environmental ethics.[14] A common reference point for this discourse is the work on ethical challenges related to the IPBES report that started in 2013 (initiated by the UN in 2010), where it was suggested as an alternative framework to the more economically oriented 'ecosystem services framework.' One of the leading theorists of this new framework is Barbara Muraca at the University of Oregon and I will mainly follow her account (Chan et al., 2018, Norton & Sanberg, 2021, 701).

On Muraca's account, relational values do not really challenge the intrinsic/instrumental framework. Instead, she sees relational values as a supplement. In Himes and Muraca (2018), for instance, they argue for a *pluralistic* account of environmental values (where relational values come in addition to instrumental and intrinsic values).[15]

To tackle what they see as a criticism from Maier and Feest (2016, 334), claiming that all values are relational, they distinguish between relational values as a process and as content. They agree that, as a process, all values are relational in that values emerge from relations (values are not locatable to the subject or the object), but that, as a content/product, we (rather) refer to what is actually evaluated and how these (relational) values are attributed and articulated (Himes & Muraca, 2018, 2). In the latter sense, relational values refer to *noninstrumental relationships*, which—on the other hand—are not grounded in the intrinsic value of the entities we relate to. Thus, relational values are a type of value 'in-between' instrumental and intrinsic values (hence, 'pluralistic account').

The more specific meaning of relational values (values as content) concerns how we refer to these values. Relational values, as something we refer to, express how certain relations contribute to a good life. The relation X is valuable for F. Valuable relations are particular relationships that are not *substitutable* (i.e., noninstrumental), which we depend on for 'leading a god life.' This is what we often call an eudaimonian account of ethics.[16] 'In recent environmental literature relational values are associated with embeddedness, collective meaning, flourishing, heritage, beauty, self-transformation, sense of place, spirituality, livelihoods, justice, conviviality, care and kinship' (Himes & Muraca, 2018, 3). Himes and Muraca also refer to empirical research underscoring that this type of relational evaluation does exist in people's minds in the way that 'people' actually distinguish between instrumental, relational and intrinsic evaluations. Additionally, they argue that it makes little sense to protect and conserve nature independently of humans (intrinsic value) for indigenous peoples, as they see themselves as embedded in a web of dependencies with the land, animals, spiritual entities, etc. (Himes & Muraca, 2018, 4). (In this way, they also seem to tackle the longstanding problem that intrinsic value seems to entail the above-mentioned idea that humans are not a part of nature.)

I have argued previously in this book and elsewhere that I think that a relational ontology is incompatible with the framework of intrinsic and instrumental values. The problem is not solved by distinguishing between 'relations as process' and 'relations as content,' because the intrinsic/instrumental set is based on a nonrelational ontology (which needs a passivist framework to be *plausible*).

When we look for goal-directedness and agency within the passivist nonrelational framework, we look for intrinsic features, but on a relational account, these features are not *there* in the sense sought for. The parts 'have agency,' but the agency is not in the part, it 'emerges in the relation.' Thus, when we leave the part-limiting view, we also become bewildered as to what we mean by 'ethical' or what it means to have value. The parts with nonintrinsic (typically instrumental) value that could figure as contrast to ethical ascriptions (intrinsic value) and give meaning to what we mean by 'ethical' are gone and the agency of, let's say, a human being cannot *easily* be distinguished from the agency of a mountain or the mushrooms in the forest. We no longer have a situation where we have a singular 'I' that finds itself limited by the ethical status of external parts. Realizing this (explicitly or not) might be the reason why we see that relational value accounts (such as the one coarsely described above) move away from the ambition to create a non-anthropocentric position and toward a more pragmatic solution ('pluralist account'), putting *human welfare* (eudaimonian) at the center.

What I mean by this is that we need to keep in mind that (extended) intrinsic value is what traditionally secures a non-anthropocentric theory of ethics. When a relational account challenges the notion of intrinsic value, it necessitates a fundamentally different response to the question: 'What makes an ethical theory non-anthropocentric?' In such cases, it may seem simpler to revert to an anthropocentric framework—such as a eudaimonian perspective—where ethical value is grounded in human flourishing and well-being. This *appears* viable and attractive because it seems reasonable to say that 'on a relational account the value of non-human parts is already entailed.'

In my view, however, this is a way to drain the potential difference that a relational account can make for very little pay-back. What is fundamentally wrong with the intrinsic/instrumental distinction is rather saved by the nuance on this account. The rationale might be that: 'After all it is better than the alternative in that it replaces the strong anthropocentric idea of eco-service with something that looks less anthropocentric.'

Is this a fair conclusion? Isn't the point rather to show that our welfare really does depend on our relations with *nature*? This *could* maybe have been the case if there was a particular reference point (e.g., possible-N relations) to appeal to in deciding which relations are really important or valuable in limiting our (welfare-probing) activity vis-à-vis nature. When there is no "possible-N reality" (or similar) in the "theoretical inventory," however, it seems that any (nonsubstitutional) relation is as good as anyone else. Relations are promiscuous. What would be the difference between (nonsubstitutional) relations established through social media or *particular* gaming characters and relations established through different activities in the nearby woods? In both cases the relations will be vital and nonsubstitutable in supporting their way of 'leading a good life'? There are no relational limits on our ethical considerations beyond relations defined by humans. I will come back to this question in the next section. At this point, I will simply lay out some common features that need to be in place for an account on relational ethics.

6.5.2 Features of a relational ethic

Despite my disagreements with the current view on 'relational values,' I agree (of course) that value neither resides in the valuer nor in the valued. But what does that mean? To get what I would call a proper hold on this question, we need to recall some of the basic assumptions in relational realism.

The general idea is that any real phenomenon occurs as dispositional activities between mutually manifesting partners. For example, water is the manifestation of the dispositional activity between hydrogen, oxygen and other partners. Importantly, the dispositions of water are not reducible to the partners at the lower level, i.e., its manifesting aspect on the relational realist account. Here I will use 'manifesting base.' The dispositions of water are determined by the partners at the interface level at which water "operates" (e.g., temperature, pressure, organisms or boats). The fact that water can quench our thirst is not a property of its manifesting base (H_2O). I use this simple example to argue that, for instance, color—often viewed as an ephemeral phenomenon by most physicists—is a real-world phenomenon that cannot be reduced to the parts constituting its manifesting base. The argument is that if something is manifested and is doing or is disposed *to do* something, it is real (on the Eleatic stranger principle). Color may scare away enemies (redness) or contribute to the sexual selection of an individual, for instance.

The same can be said about ethics, although this may be trickier. The key point is to recognize that any phenomenon occurring in the world is a relational phenomenon, as described above. It results from a relational process at the level below (manifesting X), but the way the phenomenon behaves (its dispositions), along with the restrictions given by the manifesting base, is a result of the possible dispositional partners that the phenomenon interfaces with when manifested. (As argued previously, the manifestation can be durable—an object, nondurable—an event, or cascading—a process.) If ethics is something that emerges from the mutual manifestation of parts, and it then does something (in an event), it is also real. It is real and cannot be reduced to its manifesting partners. An ethical or normative event occurs that influences someone to act in a certain way. Thus, in the case of relational realism, we are not required to argue from the outset that ethical value must be a phenomenon of a special and intangible kind taking place outside of 'what is really going on' (as would be the case where relational values are just posited without any ontological grounding). Nor must we argue that it is a real phenomenon *despite* being a relational phenomenon (cf. the notion that relational properties are extrinsic properties). Everything is a relational phenomenon.

At this point, it is important to also keep in mind that even though a disposition is not held intrinsically—'independently of everything else'—it is still possible, although challenging, to locate the different partners and dispositions involved in the manifestation.[17] In other words, I am not claiming that value occurs wherever there is a relation. The partners involved must possess the mutual capacity or disposition for 'value production,' just as the entities involved in causing 'snow to fall from the sky' must share the collective power to bring about that phenomenon. The relation between two stones lying side by side does not seem to have the mutual power to create (manifest) value.[18]

In our further reasoning, it is essential to ensure that this statement does not lead us into the opposite error, namely that partners capable of manifesting value do so randomly. It is not the case that the dispositions of water have value solely because I choose to value them. Rather, the *dispositions of water have value because of the* (historically evolved) *relations that exist.* Accordingly, we value our relationships with friends due to certain pre-established cultural (and some would say biological) bonds between people. Thus, it is the relation and the relational system that allows for the acknowledgment of certain valued relationships and makes it possible for something to be morally protected in the first place. Something does not have value *because* I value it, but because I stand in a certain relation to it from the outset.

If everything that exists is relationally constituted in the way described throughout this book, it means that we can seek out *facts* about how x is related to e, f, g, h, i and so on. Additionally, we can assert that certain relations are valuable, although this is not a fact in the same way. What is distinctive about 'relational values,' then, is that ethics comes into play to *protect* certain relations that are valued for specific reasons. It is a fact that I stand in this and that relation, *and* I want to protect it. Without concluding yet, we can already anticipate a noncontingent relationship between facts and values. If values are relational, they are *about* factual relations.

Relations can be protected in many ways: Organisms may have specific ways to protect a territory, a specific resource or their kin. The manifestation of feelings would be another example: Feelings of love and affection, anger and aggression, sympathy, sadness, shame, etc., are expressions conveying care for relationships. What we traditionally call ethical conduct is yet another type of protection. 'Ethical activity'—such as following certain rules or norms, acting out certain virtues, taking on responsibilities and obligations, and expressing symbolic gestures—is an action we take to guard certain valued relationships, so to speak.[19] When such activity is not 'properly' enacted, we say that x behaved unethically, which implies that they acted in a way that did not heed *the relation* or the instantiated relationship. A person 'notoriously full of vices,' for instance, would typically navigate life with many broken relationships.

As a manifesting partner, we can be responsible (to the extent that we have the power/disposition) for maintaining certain relations. For instance, caring for our children means caring for the relationships that are valuable to them. Relationships always depend on many partners, and in many cases, it is challenging to identify our manifesting role or responsibility. To be powerful means to 'control' a set of manifesting partners. To possess practical wisdom (*phronesis*) means that one understands how oneself and the other dispositional partners are related and how these relationships function in specific situations. As many feminist ethicists have pointed out, it is therefore not clear how we should identify what it means to act autonomously, since acting toward a goal is always a result of mutual manifestations. When we are already related and dependent, it is hard to distinguish between means and goals and, consequently, to freely choose the right means for the right goal.

In one way or another, it does not make sense to differentiate between parts that can rightfully be used as (pure) means and those that cannot, because the fundamental situation is that you are not over and against other parts. When you act out something, the act is a mutual manifestation. As a manifesting partner seeking to achieve a goal, you need to engage in a way that also follows the relational nature of the other parts. To *use* the parts means, to a certain extent, to follow them. To what extent is it meaningful to say that the use of some parts is purely instrumental if you must 'follow them' (or as previously articulated in Section 5.4, 'subordinate to them')?

However, even if we agree with this view of 'distributed agency,' it does not prescribe an ethical position. We should not argue that an ethical theory (directly) emerges from our entangled nature. Even if I am entangled and have a responsibility in this sense, it is not clear what I am responsible for. The fact that I am a co-creator of how the world is becoming (in the designated sense described above) does not tell me where to go and what to do. It indicates that I cannot begin my ethical reasoning from a disentangled position, but it does not specify what is ethically good.

6.5.3 Grounding relational ethics

Taking relations as the center of ethical concern raises the question: what makes a relation valuable and ethically significant? The main issue with relations is their promiscuity; they are neither inherently good nor bad, similar to properties. They gain weight, importance and ethical significance through something. Is this 'something' human interests, or do relations possess value beyond our interests?

As far as I can see, there are three ways to argue for a non-anthropocentric view, which resemble part-limiting, relation-limiting and whole-limiting perspectives. Moving toward a non-anthropocentric view implies the question: Do limits in nature have a bearing on ethical limits? I will now describe the three different scenarios for how a non-anthropocentric argumentation could go.

6.5.3.1 Properties (dispositions) make the relation valuable

A property-based analysis has its origin in the part-limiting trunk and we have seen that the environmental ethicists, who typically accept this as a starting point, typically search for a broader range of properties to extend the ethical realm beyond humans. I have described this approach as the opposite of a relational view, but can we find a compromise between the two? As observed, a property-based outlook characteristically seeks something with 'intrinsic *relation* an 'objective' value by referencing the properties of 'the object.' This move aims to avoid a subjective stance on the value of relations: my *relation* to X is valuable *because* X possesses the property of being rational.

I have criticized this view above. Mark Coeckelbergh also argues that property-based views are fundamentally incompatible with relational ethics (Coeckelbergh, 2012, 1.1, p. 36). This is understandable, as the property view historically emphasizes the part rather than the relation. However, one way to salvage this view is to consider properties as dispositions and view these dispositions as relational rather than intrinsic in the first place. I believe this relational dispositionalist approach could be viable because it preserves 'the *nature* of things.' But why is that important?

It seems important because, if we value relations as such, we want the relationship to function in a specific way. Without understanding or caring about the dispositional nature of the partners, we will not know how to preserve or protect the relation. For example, we need to understand the relational nature of the soil (in a particular place) to grasp what our relation to the soil 'demands' from us—if we value it relationally (Lie, 2016, chapter 5.7). This is important only if my relation to X is based on the idea that X possesses a relational *nature* independently of me (but not, of course, independently of everything else).

In Lie (2016), I argued that it does not significantly matter whether we treat X as having instrumental or intrinsic value, provided I *also* believe that X has a relational nature and that the best way to treat X is a way that aligns with this nature. If it has intrinsic value, it is *ethically right* to treat it according to its nature; if it has instrumental value, it would be *prudently right* to do so because that would be the most efficient way to treat it (see also Sections 5.5 and 4.4). The problem arises only when instrumental value *implies* that X lacks a (relational) nature, meaning it is unrelated, passive and externally determined. If this is the case, we need, as previously argued, 'intrinsic value' to secure that something can and should be treated in line with its nature (which would typically only apply to humans). A properly founded concept of naturalness on the other hand does not need intrinsic value. Treating something in accordance with its relational nature (for ethical or instrumental/prudential reasons) *can* potentially protect the relation in a way that makes the potential interests of other partners a part of human interests. This narrative however does not demonstrate that limits in nature have a bearing on ethical limits.

6.5.3.2 Where the integrated whole saturates the relation with value

In this view, the value of relations is derived from the whole to which the relation belongs. Why are relations valuable or important? From a whole-limiting perspective, this follows from the primacy of a united whole striving toward a goal. We must preserve and value certain relations because they contribute to the maintenance of the whole, ascribing moral status to the parts accordingly. Since the whole is regarded as good, failing to protect certain relations would be morally wrong.

When the whole serves as the reference point for the value of relations, it must be a unity of goal-directed, teleonomic features, such as those found in a body, a performing orchestra, a fighting military unit or a hockey team during a match. For this account to hold, the whole must exert some form of downward influence on its parts.[20] For instance, in the body, certain relations (e.g., between organs) are vital for maintaining the whole's goal-directedness, necessitating a proportionate downward influence from the body as a whole to protect valuable relations.

I have previously noted (Section 2.3) that, as humans, we do not appear to inhabit a world where all parts are embedded in a genuinely united or integrated whole.[21] Without such a whole, applying the teleological framework becomes challenging. Furthermore, even if we assume a united, goal-directed whole (as, e.g., suggested by certain interpretations of Gaia theory), it remains unclear how to establish its *goodness* without resorting to a religious foundation. (For curiosity, see Corning et al., 2023 for a broad discussion on teleonomy in evolutionarily developed living systems.)[22]

If we set aside the question of whether ecosystems genuinely exist as integrated wholes, there is broad agreement that something *can* be good or bad for an ecosystem without claiming that the system itself is good. In other words, *if* we as humans decide that it is good for this ecosystem to exist as a whole, it would also be good to protect the relational web of dependencies constituting the system.

6.5.3.3 When the ethical value of relations aligns with a specific relational order

The theory proposed here differs from most other relational theories in that it asserts that relations themselves are not empty. Empty relations "can be filled by anything" (like an artificial neural network).[23] Hence, if we are indeed relational beings but our relations are contingent, they come across as classical extrinsic properties. If relations deep down are contingent and could have been different, there is no justification for a relational ethics in the first place; they would lack significance. Therefore, if relations are empty, the theory of possible-N relations may be viewed as a means to support certain historical power interests, as post-structuralists would say.[24]

We have refrained from giving something immediate value through the intrinsic value argument, as well as any appeal to what the Greek philosophers would consider a cosmo-axiological order of the world. However, we do have a cosmology that places humans within a historically created, but specific 'relational order' of dispositional partners that exist *before* we enter as dispositional partners. We do not merely enter into relationships by virtue of leading a good life; rather, we are situated within a web of relational dependencies that enables our existence and the further pursuit of a good life. This also means that we cannot simply sit back and 'be what we are.' Since this order is an 'order of activity,' humans must be active to exist at all. The question then arises: how do we link this assertion to any ethical implications?

A good starting point would be to seek support in Knud Ejler Løgstrup's metaethical dictum: 'The good is either demanded from us or desired' (Løgstrup, 2020, 3).[25]

When the good is demanded, we are indebted to something—this is the 'Ethics of Demand.' An example of this is Kant's duty ethics and K. E. Løgstrups 'call ethics.'[26] Failing to respond to these demands implies that we are acting against what is good.

'Good,' as something we aspire ('Ethics of Aspiration'), represents the voluntary aspect of normative action. There is some kind of urge for happiness, flourishing or harmony, where we seek a goal beyond our current state. What we define as good ultimately depends on what we believe is good in itself, rather than for something else. This concept aligns with what we refer to as teleological ethics (i.e., eudaimonian and utilitarian ethics).[27]

6.5.3.3.1 'ETHICS OF DEMAND'

When we say that something is demanded from us, we may only describe (as best as we can) what it takes to achieve a goal: 'If you want to build a civilization on Mars, here is what is required.' This is what Immanuel Kant called a hypothetical imperative, which does not (at the outset) imply any ethical consideration. In contrast, when something is demanded from us in an ethical sense, the *demand* does not originate from our free choice of the goal. Kant famously argued that we cannot freely choose a goal that implies lying to or killing another person without committing an immoral act. There is a limit to how we can treat others, as some individuals have intrinsic value and must therefore be treated as such. As we are now considering a situation where we already assume that we are placed within a relational order, this demand does not stem from the status of the other party but from relations themselves. In the case where the requirement originates from a relation, Løgstrup would call this a third type of ethical theory: The ontological tradition.

Løgstrup argues that this type of demand does not depend on what we feel or what we think, but originates from a prelingual bond that instantly calls us to act in certain ways. Løgstrup argues that these bonds are universal (among humans). They are not voluntary or contractual but rather silent pre-ethical calls rooted in the fragility of human interdependence. Trust, first and foremost, along with mercy (compassion), openness of speech and courage, are *life expressions* that arise spontaneously from the *factual* or ontological interdependence between humans. They occur before any rational discussion about our ethical conduct can begin (Løgstrup, 2020, chapter 2).

We may view these expressions as manifestations of our *fundamental* relatedness, according to Løgstrup, and something is amiss when these expressions (dispositions) are not (spontaneously) manifested in the situations where they should occur. They are expressions of a relationship, not specific properties of the subject or object. Individuals who act as if these silent calls do not matter are acting unethically.

The view that we explore here could rightfully be called ontological, but we want this view to explicitly include more than humans. When that is the case, the requirement that Løgstrup imposes might be too strict. Not all relations that we want to include qualify as silent calls. Some of the relations might have this quality, but not all. What we want to explore here is not pre-ethical relations that call us to ethical action. Our aim is more generally to ask: what kind of difference does it make to view relations as the starting-point for ethical concern?

If this is what we are searching for, what seems to be a general feature is 'balance.'[28]

For example, if I borrow something from someone, I feel an obligation to return it. If someone trusts me, I feel obliged not to break that trust. If something is afforded or given to me, I am compelled to reciprocate. This implies that certain relationships require specific actions to maintain the *balance*. The key point for us is that the call comes from the relation, not from the other partner per se. Relationships and our perception of them require a certain balance, and acting out of balance is central to ethical misdeeds.

To move on further within the frames of the 'ethics of demand' account, let's compare this view with what we already seem to accept in the case of Kantian reasoning. If we are not ready to accept the role of ontologically stated claims about relations, we should not do so in the case of Kant either. Why should we treat other people as goals in themselves (according to Kant)? This claim originates from an *ontological claim* about the part that we call humans. What we are referring to a description of what humans (really) *are*. The transcendental features of experience are *descriptions* of human rationality.[29] From this fact (is) he moves on to argue that humans have a duty to exercise this faculty (ought), hence to ask ourselves whether our maxims can be universalized. (i.e., the very exercise of human rationality). If the universalization leads to a contradiction, I should not act in line with the maxim (i.e., 'lie to a person to save myself from an awkward situation'). In the case of a relational framework, I have a particular descriptive understanding of the relationship (is) with another human being that demands me (ought) to act in specific ways. Thus, rationality does not discover what is morally right; it identifies what kind of acts would contradict the nature of the relation.[30]

What this illustrates is that it is challenging to justify why we should act morally toward something else without seeking to understand what the relation *is*. Thus, the key to finding the right ethic (within this meta-ethical framework) comprises two components. First, we assume that a relation can be described with a reasonable degree of accuracy. Second, it is important to recognize that the relation does not directly dictate the right action. As humans, we stand in a relation to the relation. We interpret the relation or act based on a certain presumption of it, which means that we enter into a particular relationship based on the relation.

Taking the first aspect as our starting point, we can view injustices, both past and present, as manifestations of a misalignment in the point of balance that originates from incorrect descriptions of the relation. For instance, the relation between a master and a slave, or the relation between humans and allohumans. I might feel the call of duty to maintain a balanced range based on my learned and accepted understanding of my relationship with X, while, upon reflection, that underlying relation should have been understood and described differently. I may grow up in a society with a rigid hierarchical structure, for example, one that demands certain behaviors to maintain the balance of the prescribed relationships and penalizes actions that do not honor these dynamics. In a society where men are positioned above women, for instance, what does that entail? It means that there is a (culturally interpreted) balance point that must be acknowledged to sustain the described relation.

If 'might makes right,' it follows that relations are contingent. Whoever gets into power is free to describe them in whatever way they choose, tacitly deeming these descriptions as mere tools for the execution of (pure) power. Acting "morally" correct is simply acting instrumentally right according to the way those in power have defined and characterized the relation. If there is a connection between 'relations and ethics,' relations cannot be contingent however. If relationships are random and contingent, the (proposed) balance point would only grant me entitlement in a purely instrumental sense. If I bend my neck for Caesar to hit the right balance point, it is only because he had the *power* to describe the relation, not because I agree with how the relation is defined. Had I agreed (naively or genuinely) to this, my bow would not only be instrumental, and Caesar would have nothing to fear from his subordinate(s).

The alternative perspective would be to assume that it is possible to be right or wrong in how we describe and understand our relatedness and dependencies. Certain descriptions are better than others. There exists a specific understanding of the relation between rulers and the ruled, between men and women, between parents and children, between humans and other animals, between fungi and the roots of a tree, and between parts and wholes. Moreover, in this framework, we assert that getting the description right brings us closer to understanding what is specifically demanded.

This is where relational realism reappears. First, we assert that we are relational beings and that our relatedness is not contingent. Second, we propose a possible account of the relational order (possible-N), suggesting a framework for how a correct description would appear (with exceptions for a number of social relations).

The second aspect concerns how we choose to act out and interpret relationships. Even if we were able to accurately describe the relation, it does not fully prescribe what actions are necessary to achieve balance and protect that (specific) relationship. We may agree on the description but disagree on its implications. The statement that women and men 'are equal' does not dictate how we should act in accordance with that equality. A society may develop norms and role descriptions that limit us in certain ways, but such norms can still be questioned or deemed incorrect despite having the same understanding of the relation. More importantly, the right description does not determine whether the relationship is worth protecting. This is true in a very specific sense, as I can decide that nonexistence is preferable to existence. However, if existence is deemed better, I cannot choose to remain inactive or abstain from participating in the activities that make existence possible. Thus, our understanding of the relations matters to (but does not alone determine) the question of which relationships are worth protecting and how this should be accomplished.

From what I can gather, we may identify three types of relations.

First, we have relations that are radically asymmetric. Asymmetric relations are those in which we cannot reciprocate. An example of an asymmetric relation is the description of our relationship to God in the New Testament. God's sacrifice of His son, along with Jesus's self-sacrifice, imposes an enduring debt on humanity—one that can never be fully repaid or reciprocated. Our existence as humans seems entirely dependent on God. Our relationship with God is inescapable. If the Bible accurately describes our relation to God (and Jesus) correctly, we are (under certain interpretations) called to follow God's words in the New Testament to act ethically. The overarching narrative of the New Testament exemplifies an asymmetric relation of radical dependence. One may question whether 'God sacrificing His son' is the appropriate "proxy" for this relation, or whether it is a typical misplacement of the dispositional partners involved.

If we believe that the proxy is incorrect, it does not mean we should dismiss this *type of* relation outright. It is a fact of life that we have not created our own lives; we depend on a wide range of relational partners—past and present—that we can never reciprocate. Existentialists might argue that there is an 'eternal weight' on our shoulders, and when we are not burdened by it, we sometimes experience the opposite: the magic of being 'immersed in the orchestra' playing its own songs, harmonious or not. We feel blessed, even though we depend on the orchestra and the orchestra does not strictly depend on us. This type of relation is not chosen, which means that the onus is on us. To protect the relation, we must act toward a balance point without ever being able to achieve it. A typical way to do this is by expressing gratitude, given that we choose existence over nonexistence.

One could argue that this conveys a certain direction to life. If all relationships were symmetric, there would be no direction to our lives according to the ethics of demand. It is only the gap between what is demanded and what we can do that creates a sense of direction in life. If you are indebted to no one or nothing (autarkic), you can go anywhere, and there would be no ethical (only strategic) limits to your actions. A directed act, as intended here, is one that aims for 'redemption'; it seeks to find the right compromise in a given situation, knowing that balance may never be struck.

The challenge with radically asymmetric relationships lies in the nonspecificity of your counterpart. What you owe is not a unified phenomenon. Your parents, your kin, your ecological kins, your clan, your society and the lineage of evolutionary predecessors who 'sacrificed' their lives for your present existence—all these relational partners that have made you possible (rendering the distinction between adversary and ally meaningless) are candidates for inclusion.

God is a preferred counterpart for many. Since there is no proper unity that can serve as the correct counterpart, I would suggest considering possible-N. Possible-N traces back 13.8 billion years and encompasses the almost miraculous development of life and consciousness. It represents the fundamental conditions that have made me possible and that afford further possibilities. The iterations (possible-I) can be questioned, as they are to some extent a result of human choice. Some will praise them, while others will not, so they do not qualify as a radical asymmetric debt. Possible-N on the other hand 'is just there.'

What this (tentative) suggestion implies is that possible-N represents not only a factual and prudential starting point for our further transgressions but also something we are asymmetrically indebted to. This provides our actions with direction, obliging us to strive for a balance point that generally leads to solutions beneficial to our possible-N partners. On this account, we can never reach a point where, for example, our farming practices perfectly align with possible-N, but we are nevertheless morally directed to seek the best option. We are *generally* obliged to understand the irreducible relational entanglements and make compromises that do not treat human needs and preferences (the marked) as unquestionable starting points. Just as Kant would argue that we cannot claim that we 'need to lie,' we cannot assert that we 'need more energy' and arrive at a prudent compromise (symmetry between possible-N and possible-I) that accepts the need as a fact. We are indebted to a noncontingent reality that limits our moral aspirations. A challenge here is obviously that it is hard to pin down what protecting this asymmetric relation means in practice. (But as previously stated, we are not searching for an account of moral status.)

A second type of relation is symmetrical relations. These are relations where we can choose the specific partners in the relationship, but we cannot choose the demands that arise from it. If I wish to challenge the demands, the burden is on me. There is a certain order to the world that is not chosen. Simple examples include friendship, acquiring a pet or being employed. More complex examples involve our interactions with nature. Do we fail to meet the demands if we do not consider the entire 'package deal' when we seek to change certain manifestations (cf. nexus efficiency)? If we do not accept a radical indebtedness to possible-N, we can nevertheless accept that our containers should be symmetric. It might however be difficult to assert that this is a moral demand.

In general, in symmetrical relations, it is possible for me to reciprocate, but we also need to understand and act in specific ways to protect the relationship. In many, though not all, cases, failing to do so will call forward ethical repercussions.

A third type of relation is 'no-demand relations,' The relationships we enter—freely or otherwise—carry rules and entanglements that are arbitrary. If someone wants to prevent me from doing something, the onus is on those who want to stop me. They must demonstrate and convince me that the relationship I enter demands what is required of me. This is what classical liberalists consider a general starting point for human interaction. In reality, these are relationships with rules that 'one follows' and are often the primary target of generational revolts.[31] As such, they are not morally binding but may nonetheless serve to protect unjustified and arbitrary power relations, indicating a misuse of what lies at the center of the 'ethics of demand.' If there is a misuse, one is justified in revolting based on a liberalist stance.

These three types of relations are not clear-cut and often intersect, but within this 'ethics of demand' framework, they are also confused. We should not for instance act as if we are in a symmetrical relation if it is not symmetrical, and not all symmetrical relationships are social. Not all asymmetric relationships can be made symmetrical (e.g., the quest for immortality), and 'no-demand relations' are less abundant than we typically assume today. Throughout this book, I have expressed my views on how these relationships are confused and how they might be more accurately perceived, so I will not revisit this here. My point is that getting the descriptions of these relations right matters because it enables us to better understand what they demand from us. Whether we uphold this demand through ethics or some form of prudent hypothetical imperatives depends on culturally and historically rooted decisions.

6.5.3.3.2 'ASPIRATIONAL ETHICS'

This view seems to align more closely with a relational realist perspective. We are in pursuit of something—a good life—and we recognize that our relationships must be protected to move forward. The drive to act ethically does not stem from a void (as in an asymmetric relationship) but arises from a desire to achieve a better future. This perspective is supported and deepened by current relational value theories (see Section 6.5.1). However, I have argued that this account is problematic because the ethical status of relationships is based on specific ideas of what it means to lead a good life. For example, it is wrong to cut down the Amazon rainforest if a good life depends on the forest's integrity. This issue becomes even more complex when we consider that we may not reach a consensus on what constitutes a good life, particularly within a liberal political framework.

Consider again the planetary boundaries established by the Stockholm Resilience Centre, which asserts that there are nine boundaries we cannot exceed. One person may argue, however, that he wants to lead a good life focused on 'enjoying the moment,' while another might claim that exceeding these boundaries would lead to the collapse of modern civilization, which they view as a positive outcome. An aspirational account of relational ethics, therefore, is one where relationships are ethically significant only to the extent that they support our desired ways of living. In other words, relations are empty at the outset.

But the question here is not how relations become important when certain goals are in place, but what it means for an aspirational ethic if we already assume that we are relational beings. The critical question is what this account implies if our aspirations must align with a specific relational order. If all humans are relational beings dependent on each other and a specific, noncontingent natural order, it follows that our aspirations cannot be freely chosen. Any aspirational goal setting would likely lead to fairly similar conclusions about which relationships should be safeguarded by ethical limitations. In other words, I cannot simply claim a preference for 'enjoying the moment' and freely pick out unrelated sets of dependencies. While we may choose our aspirations freely, this choice can only be justified if we also acknowledge the possible-N limits that apply to everyone. The ethical value of the noncontingent order follows from what we consider a life worth living, but 'a life worth living' does not specify which dependencies should be protected by ethical constraints. This creates a rather weak link between natural limits and our ethical obligations. There may be room for further refinement here, but as it stands, it is challenging to connect how possible-N might restrict the liberal buffet of life projects. Is there a way to make this connection closer?

Arne Næss' ecosophy T represents a position that might be helpful to illustrate a closer connection between limits in nature and aspirational ethics. In his book *Ecology, Community, and Lifestyle*, Næss argues that normative statements cannot be derived from other premises (in line with the is/ought distinction) but must simply be asserted (Næss, 1989, 68 ff.). The highest norm, according to Næss, is self-realization (Næss, 1989, 84, 164 and 197). What we all should aspire to is 'self-realization.' Ecosophy T also entails a relational ontology however (Næss, 1989, 55). If we assume self-realization to be the highest norm and recognize that we are 'dots in a relational field,' then 'myself' would be more fully realized if my pursuit of self-realization also includes the self-realization of other dots in the web (Næss, 1989, 199). The argument is simply that 'myself' already includes these other dots as a matter of fact; hence, seeking self-realization in isolation would be a deprivation. (Næss, 1989, 173–174).

Næss argues that this position does not primarily stem from rational consideration but from identification. Identification is also a realization in the sense of 'discovery.' The more I identify with other "dots" (i.e., partners), the more truthful my identity becomes (Næss, 1989, 175). The implication is that there is no genuine conflict between my self-realization (flourishing) and the flourishing of what I identify with. If someone destroys or threatens a forest with which I identify, they are also partly destroying me, just as killing my brother or child would also 'kill me' in a certain sense. This is why Næss emphasizes the role of feelings and seeks to clarify what 'identification through feelings' entails, noting that such identification goes beyond superficial attributes (e.g., 'cats are cute, fish are stupid') (Næss, 2002). This theory is what Næss refers to as ecosophy T.

But the problem described above re-emerges, as it seems to be the case that we can be a part of and identify with nearly anything. We may feel upset or even go to war when someone threatens our nation or attacks our favorite band. I can strongly identify with people I have never met (e.g., celebrities) or characters in video games. The fact that we identify and extend our relational selves does not necessarily imply alignment between our aspirations and a specific (noncontingent) relational order; thus, there may be no particular connection between ethics and possible-N from this perspective.

An answer to this issue is to suggest a distinction between what we can identify with and what we should identify with. This would necessitate a further normative criterion, such as: 'we should identify with our real relational partners.' Identifying with partners that are genuinely part of me, with whom I may not have yet identified, would make flourishing a project of becoming real. Conversely, identifying with partners that are 'not really a part of me' would render my flourishing hollow. The question then becomes: What am I truly a part of? (see also, e.g., Mathews, 1991, 135 ff.). If I were imprisoned, my everyday manifesting partners would typically be other prisoners, guards, closed doors and walls. And what about my modern-day partners? Would I not become more real if I identified with the flourishing of my Samsung phone, my car, Patagonia jacket, the Minnesota Wild or Norway as a state? Is there a case for authentic self-realization?

As mentioned earlier, I do not believe we can reach a consensus on a correct description of our relational nature. However, we are closer to addressing what we should identify with if we agree that this is a crucial question that every society should continue to discuss, even though the question may never be settled. Failing to engage in this dialogue risks ceding our relational nature to prevailing power interests, which may either ossify unwarranted descriptions (of relations) or persuade us that relations are empty and should be filled with whatever our part-limited selves find preferable in the marketplace.

At first glance, Næss' ecosophy T appears to provide a useful account of how possible-N limits and (aspirational) ethics can combine. If self-realization is the ultimate norm and it is a fact that we are relational beings, it follows that my self-realization depends on the self-realization of my partners. If I act in ways that restrict the self-realization of my manifesting partners, I am acting against what is ethically right. Moreover, if I extend myself by identifying with my real partners, I become more real as I identify. Therefore, the limits of what my possible-N partners *can do*, do *prima facie* impose ethical constraints on my actions.

However, it still remains unclear what it means for me to self-realize. Initially, Næss advocates for a pluralistic and liberal stance. The more diversified the better (Næss, 1989, 200 ff.). The deep ecological movement is consistent with a wide variety of human flourishing and ecosophies (Næss, 1973). This, however, seems to reintroduce the issue that relations are contingent on our goals.

I have now outlined various perspectives on how limits in nature, understood as relational limits, might have a bearing on our ethical conduct. As previously noted, the reader should view this as an exploratory attempt, with proper detail and broader references omitted. The promise of these perspectives may depend on viewpoints and criticisms that I have yet to consider.

6.6 Limits in the economy: The self-perpetuation and self-limitation of economic growth

Throughout this book, I have argued that possible-N represents real limits in nature. To what degree are these limits reflected in our economy?

First, we need to recall that possible-N dispositions are those given for free. We must sometimes work to *manifest* these dispositions (e.g., collecting water, hunting prey, making fire, and picking berries), but the dispositions themselves come at no cost. If we want to transcend what is offered for free, we must establish and maintain possible-I *arrangements*. Although the distinction between what is free and what is not is not strict—and never will be—this is of minor importance here. The general point is that possible-I arrangements require thermodynamic work and cooperation, which ultimately create value, including transactional values: items and/or services that can be offered to others for compensation.[32] Thus, as a general rule, we can say that going beyond possible-N is what generates profit, products and jobs. Initially, we would not consider the presence of air (for breathing) to be a product that can be sold on the market. Therefore, in a society where the goal is to create jobs, profit, and ultimately economic growth, it is generally the case that we can only achieve this by transcending possible-N and variously undermine the web of possible-N relations. Consequently, a product or service sold on the market carries the burden of possible-N transgressions. The more we transgress, the better the economy, provided that the transgressions yield returns greater than the added work. Thus, the destruction of possible-N webs can also benefit the economy to the extent that it creates needs that can be met through the provision of compensating products and services. If this is the case, we have a self-perpetuating system that, as a result, bends toward the destruction of the possible-N web.

What is economically possible (possible-E) is clearly broader than possible-N, and what is technologically possible (possible-T) is broader than possible-E. Similarly, what is scientifically possible (possible-S) surpasses possible-T, and what is logically possible (possible-L) extends beyond possible-S. Possible-E is predicated on the available organized resources (such as manpower, energy and capital) needed to perform a specific activity at a given point in time (see also Hagens, 2020, pp.4–5). What we identify as scientifically possible may not be realizable technologically, and what we consider technologically viable could be far from profitable.

However, when we introduce *economic growth* into the equation, it becomes challenging to track the relationships among possible-N, possible-T, possible-S and possible-E.

Firstly, what is economically possible refers to what is profitable, rather than what can be deemed as economically feasible in relation to possible-N. From a societal perspective, a primary goal of the economy is to organize human activity in a manner that meets human needs and allocates resources to efficiently create value for further growth. Without a financial surplus, there will be nothing left for reinvestment to drive that growth. Entities that cannot reinvest and generate revenue (or a prospect of revenue) for investors will struggle to survive in a capitalist economy. A decline in revenue will reduce the capacity to support human needs or preferences. While it was obviously economically possible to build the pyramids, such an endeavor would not be considered 'economically possible' today, as it would likely not yield returns for future investment.

Secondly, the determination of what is economically possible now (i.e., profitable) can be deferred into the future. Companies may raise capital now with expectations of future returns, and nation-states may issue bonds anticipating future financial returns. From this perspective, possible-E might be even broader than possible-T at a specific point in time. While the belief in a technological breakthrough may be unfounded in some cases, it can be validated when growth is factored in.

More generally, we may encounter a situation where debt creates growth (and immediate financial returns), but where the debt (financial power/possible-E) grows at a rate that exceeds the growth rate of the economy (GDP) or its potential growth rate (Hagens, 2020, chapter 6.5). In such cases, it becomes challenging to maintain the idea that possible-E represents the 'realistic version' of possible-T.

Furthermore, if we define growth as 'eternal growth,' the distinctions between possible-N, possible-E, possible-T and possible-S become even more blurred.[33]

Economic growth carries the promise of transcending limits that the current economic situation does not permit. This suggests that when we consider *limitless* growth, we do not truly find the limits of the economy within the economy itself, but rather in possible-T. With *eternal* growth, there is, in principle, no limit that cannot be surpassed, except for those limitations that we are unable to overcome through technological advancements. This means that what is possible-E theoretically aligns with what is possible-T.

However, what are we unable to change technologically? With eternal economic growth, it remains possible to even exceed what is technologically feasible at any given moment. Consequently, the ultimate limit to our economy resides in the realm of science. The economic possibilities are theoretically determined by what is deemed profitable at a specific time, but if we assume eternal economic growth, the limits of the economy will ultimately reflect the limits of scientific advancement. This underscores the paramount importance of science in our modern economy and in determining what is seen to be the limits of nature (by science).

So, we need to ask: what is science if not the science of nature?[34] On the account presented in this book, science investigates what is possible-I in relation to possible-N. However, is possible-N actually evident in science? We seem to be back to the issue of possible-N and the disagreement surrounding its definition, or even whether anything can be labeled as possible-N at all. I have argued, and perhaps more convincingly in Lie 2016, that science does not accept the concept of naturalness. The only limits that science can ontologically accept are those pertaining to possible-I, which are presumably too ossified to be 'moved by humans' (often deemed to be laws and constants). Science does not distinguish between truths found in possible-N and possible-I (see Section 3.3). This is particularly concerning if I am correct in claiming that the 'eternal growth assumption' ties the limits of the economy to the limits of science.

As we have seen, possible-I truths have a wider scope of (relational) dispositions, which we investigate under contrived conditions, most prominently in the laboratory. I have also argued that in the current landscape of scientific discovery, one will need an essentialist account if one denies the distinction between possible-N and possible-I.[35] The idea is that each part has an eternal essence, contributing *only* its essential properties. To uncover the essence of each part, one must screen off (cf. *ceteris* paribus) all other parts, which is what the (illusion of the) laboratory "enables." The *relation* between the parts is secondary (monocausal), and there is no reciprocity between the parts.[36] There is no history—just parts with contingent essences that happen to come together and produce a certain outcome. Parts are not limited by their relatedness but by their essence. Thus, one can map all the parts separately and add or subtract them at one's own discretion. This may seem like a caricature, but overcoming the truths of a caricature requires more than just minor adjustments to one's perspective (e.g., by discussing "multicausality"). If the scientist does not operate with, or believe in, possible-N, there is no reason for the economist or the engineer to even consider the asymmetry between possible-I and possible-N.

This situation is also quite fortunate for those who believe in perpetual economic growth for two reasons.

To the extent that we do not recognize that limits are historically defined and relationally constituted (and therefore cannot be manipulated in isolation), we tend to overestimate the value of the possible-I container. Efficiency, and thus profitability, are often assessed solely based on the possible-I container. However, true efficiency, when considering the costs of work and maintenance, along with the ramifications of known side effects and the magnitude of both known unknowns and unknown unknowns, is only arbitrarily accounted for (viz. 'agnotology'). The more minimal the container, the more profitable it appears. A 'minimal container' is, as previously argued, one that safeguards its primary function from possible-N dispositions that threaten to compromise that function. The ability to withstand such threats is often equated with product quality. However, the container is actually part of a larger system. The more the container are considered vis a vis this larger system of possible-N partners, the less profitable it becomes, ultimately limiting the range of possible-E and hindering economic growth (at least in the short term). Thus, capitalist growth rewards modern essentialist ontologies, and vice versa. It *is profitable* to assume that each container is constructed from parts where each part performs its essential function and nothing more. If these parts are *in reality doing* more (as will always be the case from a relational perspective), additional liability will arise from exercising *the property rights of the* container. For example, if glyphosate has properties that, when combined with other possible-N relational partners, contribute to health problems in human populations, the liability for those who own the technology will be greater, making it less efficient and profitable. Contested areas of science typically involve doubt and causal complexity, especially when property rights come with (unprofitable) liability.

Thus, the success of capitalist growth largely stems from its ability to "explain away" the effects of products that extend beyond the accountability of the isolated possible-I container. If the limits of nature were inherent to the container, they would manifest in possible-E through higher prices, which would hinder consumption and, consequently limit possible-E. In many cases, this would reveal that the 'new invention' *is not necessarily more efficient than its* predecessor. Symmetric possible-I containers (which today also include the higher-order technosphere infrastructure—facing its own asymmetry problems) do not perform as well as their asymmetric counterparts. They do not generate profits for reinvestment in the same way that asymmetric containers do. Therefore, economic growth is not inherently problematic in relation to environmental problems; the issue arises when economic growth is driven by asymmetric containers (that lead to overshoot).[37]

The notion that science can define the limits of nature as fixed is a dangerous idea. It is perilous because science commands authority, and economic considerations significantly influence our modern societies. Eternal growth uses possible-S as its reference point for natural limits, informing economic theories that perceive the world as a collection of randomly organized parts that can be manipulated as if they possess essences. This perspective legitimizes possible-I–based nexus-efficiency and the (principal) right to evade responsibility for externalities.[38] This skepticism is why I view the scientific enterprise with caution today. Science is not a neutral institution in our modern society; rather, it is deeply entwined with the project of (asymmetric) eternal growth.

However, if science could establish an ontological distinction between possible-I and possible-N, it would clarify *what it truly takes* to achieve genuine efficiency and real growth in the economy. This is a call for optimism, because there seems to be no realistic way to end growth without also totally collapsing the economy. If we could progressively accept only symmetric containers, we could slow down the economy while simultaneously using our minds and "industrious" attitude to find ways to optimize limit-arrangements. But this would be hard to achieve if we are not also simultaneously able to accept and recognize that fleedom does not deliver on its promises.

The question of economic growth was what led me to study philosophy in my early 20s. Why does our economic system necessitate growth? What drives economic growth? Is it possible to end growth without societal collapse? Why is growth a threat to the environment? Why do these questions not emerge as central issues in public debate? I have not seen these questions receive satisfactory answers, especially when considered together.

Without attempting to provide a definitive answer myself, I believe that some of what has been discussed earlier in this book makes it easier to identify the major driving force of the system. Not surprisingly, I think we need to seek our answer in the consumer aspect of the equation and the concept of freedom that drives consumption.[39]

What is economic growth? Economic growth is the process of rearranging the components of our economy to align with our goals. We seek unity and alignment with the world *through* the reorganization of these components in space and time. This process contains a built-in contradiction that drives growth. Modern consumption calls for detachment, disentanglement and independence, yet what we truly seek is unity in terms of ready-to-hand interaction with our environment (see Chapter 3.7). Growth operates as a machine capable of continuously providing new products that seemingly alleviate our dependencies. New products and the expanding commodification of previously noncommodified items enter the marketplace because they fit perfectly with fleedom. Since this is a demand on the world that can never be fully satisfied (c.f. trivial optimums), growth has no limits.

If this is accurate, the system we have constructed is ingenious.

By disentangling the world, we not only create monetary values through this process, but we also generate monetary value by producing connective spaces (or connective problems) that arise from the disentanglement, reflecting the relational nature of existence. The more your garden aligns with your *specific* preferences, the more "gardeners" you need to maintain it. The more we disentangle and damage our ecosystems, the more 'connective-services,' 'connecting-instruments' and 'connecting-bureaucracies' we need.[40] The repair (so to speak) and the maintenance of potential partnerships manifest as growth in GDP. Modern capitalist societies can only thrive when organized,[41] but as previously argued, the need for organizational structures is not the result of ideology or intent; it stems from the relational nature of the world. The more we dissect the world, the more problems we create, which can only be addressed through further growth and complexity.[42] Picking the world apart has its price. To pursue or forgo economic growth is not merely a political choice. Growth is driven by consumption rooted in a specific conception of freedom. Therefore, we cannot address the issue of economic growth without reexamining our fundamental understanding of freedom.

For this reason, the problem of economic growth concerns not only those who worry about environmental issues but also those who are concerned about 'diminishing returns.' Often, these are the individuals who advocate explicitly for growth yet end up blaming ideological intentions for the increasing complexities of politics and bureaucracy (cf. Peter Thiel). A libertarian state is proposed as the solution. However, a (techno-scientific) libertarian state inherently assumes that each component is separate and can be recomposed at our discretion. This is precisely the mindset that has contributed to the creation of the whole-limiting technosphere that operates behind our backs. The core issue lies in the asymmetry between possible-I and possible-N. If we truly understood what it takes to create symmetric containers, we would also grasp the actual efficiency and real value of such containers. The phenomenon of 'diminishing returns' would become immediately apparent, rather than manifesting as a mysterious stagnation that is often explained by various semi-conspiratorial self-serving theories. (Even theories about antichrist).

We could, of course, inject more energy into the system to maintain and prolong our current state, but this would not alleviate the non-chosen whole-limiting pressures—in fact, it would likely exacerbate them. Alternatively, we could embrace a more unrestrained approach by moving away from the Promethean choosing-controlling mindset but this would also challenge the part-limiting concept of freedom. The goal was to achieve independence, not to become part of a system and a process that no one controls. Some transhumanists might advocate for a more untamed and wild frontier culture by referencing the cultural (i.e., postmodern) critique of modernity's subject: 'We were never modern anyway.' While this critique may point to a relational understanding of the subject, it fails to acknowledge the interconnected nature of the world. The untamed will eventually be tamed and constrained by dependencies that, under this regime of wild frontier culture, will be even less predictable. Who would genuinely desire increasingly harsh environmental conditions combined with a whole-limiting technosphere that does not *even* support the part-limiting concept of freedom?

Notes

1 The answer to the question about Martian life clearly depends on what we mean by 'resolved,' however. Whether by 'resolved' we mean that humans can survive there, or if we refer to the success of making the planet 'habitable'—that is, creating a magnetic field and an atmosphere conducive to normal breathing, terraforming the planet to provide food and generating sufficient energy—it is still safe to assert that these outcomes will not occur spontaneously. Most of the systems required will rely on human decision-making, many of which could have dire consequences for all or a significant proportion of the inhabitants (see also Torres, 2022, chapter 2).

2 These notions (perfectionism, communitarianism, supporters of a positive concept of freedom and republicanism) are viewed as challengers to classical liberalism, as they argue that our autonomy depends on our ability to *actually* pursue our own goals. For instance, the state should provide a common education system. The problem (says the liberalist) is that 'actually pursue' entails that we agree on what we *all* should pursue. This is a central issue that I will come back to in different ways, and I will mostly use the term perfectionism to designate this position from this point forward. I refer to classical liberalism as neutralism.

3 Stephen Toulmin argues that the nation-state constitutes a new whole-limiting order. This order is primarily a social construct rather than a natural order. To the extent that nature comes into the picture, it is not in terms of a whole-limiting order. It is rather the view that science portrays nature as passive, transparent and controllable. This perspective separates the normative sphere from nature, aligning with the subject/object distinction that underpins scientific objectivity, ultimately focusing solely on the human sphere. The whole-limiting order is, as previously discussed, not predetermined but rather a chosen construct. This choice becomes clearer as the nation-state evolves, disentangling from religious beliefs (theocracy), as the middle-class bourgeoisie gains power and the aristocracy loses its legitimate standing due to a lack of rational legitimacy. Toulmin posits that the rationalism of the Cartesian-Newtonian worldview gains traction as a common ground, particularly in the aftermath of the exhausting conflict between Catholics and Protestants (the Thirty Years' War). This conflict provides a (natural) framework for the nation-state, diminishing the significance of emotions, will and (diverging) anthropologies (including religious orientations), which came to be viewed as sources of conflict and turmoil (Toulmin, 1990, chapter 3).

4 Hobbes' idea of the 'state of nature' is understood as a scenario where there is no (nation)state, only individuals in conflict with one another. This idea, however, only yields meaning within a part-limiting view. In a whole-limiting view, there would be a *natural order* where, e.g., every part has its natural place, and where the parts do not have natural rights (birth rights) to do what they want but rather have rights according to their place. Therefore, the radical aspect of Hobbes' political theory is not merely the novel concept of the state's role or his (semi-authoritarian) social

contract, but rather the outright rejection of a *natural whole-limiting order*. The assertion that the state of nature embodies chaos and contingency is essential for recognizing any organized society as a human construct. Consequently, the more or less organized states that preceded Hobbes can also be seen as contingent artificial constructions, where those in power have persuaded the rest of society that they are part of a natural (whole-limiting) order. This perspective underscores the importance of acknowledging that any order 'above the individual' is a construct, implying that the only true entities in society are individuals, who are in constant conflict with one another but can coexist if properly managed. If this is our foundational view of the state, it follows that liberalism (later on) becomes the default 'natural' position, overshadowing all other perspectives.

5 Natural rights are intended to secure our autonomy and are said to 'have no content.' Without these fundamental, formal rights, I cannot pursue my own goals autonomously. This is why the state must protect these rights as fundamental rights. However, the specific goals I pursue should not concern anyone else or the state, provided they do not infringe on others' rights to do the same. The right to free speech does not dictate the content of my speech; what I choose to express is (within certain limits) my own business. As mentioned in Section 2.4, humans are considered the only creatures capable of acting independently of 'external causes' and in alignment with an intentional goal. The individual right to 'lead a good life' encompasses the right to define this goal for oneself.

6 Famously, Isaiah Berlin argues that a state that legitimizes its power through positive rights—where a specific idea of what it means to be human and to lead a good life is upheld—will become authoritarian, or, in the terminology used in this book, 'whole-limiting.'

7 Whether the mammoth really can be resurrected remains a debated question (e.g., Zimmer, 2013).

8 The symmetry requirement between responsibility and ethical status is most notably expressed by Kant: we are only ethically liable for trespassing limits concerning parts that can act freely through their rational capacities (Kant, 2015, § 5–6 and 1990, third antinomy). We cannot be held morally responsible for transgressions of limits affecting allohuman parts. Paul Taylor, Baird Callicott and Holmes Rolston, in particular, argue that we need to distinguish between what Callicott calls *source* and locus. (They all have their own vocabulary to express the same view) (Lee, 2003).

9 Expanding the realm of morally considerable parts is often referred to as extensionism (Carter, 2007, 26).

10 O'Neill distinguishes between the following meanings/interpretations: (1) noninstrumental value, (2) nonrelational value and (3) objective value. The paper investigates how these meanings are conflated, suggesting that we can have, for example, noninstrumental value without objective value, and so on.

11 Longer back in time 'relations' have played an important role in feminist ethics, but as a part of a bigger picture where especially the 'ethics of care' has played the main role (e.g., Noddings, 1984).

12 From now on, I will call the ethics based on the intrinsicness-ontology property-based ethics in line with Coeckelbergh, 2012, 36.

13 The difference in narrative seems to be what motivates Patric Baard to argue that the problems that are addressed by those in favor of relational values are already 'dealt with' in the environmental ethics discourse (Baard, 2024).

14 See especially the varieties of contributions to *Special issue in environment and sustainability: Sustainability Challenges: Relational Values*. Ed. Pascual et al. (2018).

15 This is in line with how relational values are mostly rubricated, i.e., relational values do not challenge the traditional intrinsic/instrumental scheme but fill a gap or come in addition (Norton & Sanberg, 2021).
16 We often distinguish between ethics that defines the good in terms of what we should seek as being (ultimately) good or what is right (duty ethics). Utilitarians are also defining their ethical position in virtue of the ultimate good, but are not called audiamonian, as 'ultimately good' is defined very differently. Both of them fall under the umbrella 'teleological ethics,' however. Eudaimonia (well-being) is the expression that Aristotle uses for what he finds to be the end-goal of human life. The capability-approach (and revived virtue ethics) is often associated with eudaimonian ethics today. In Muraca, 2016, the reader will find that she has a more nuanced view where she distinguishes between fundamental relational values and audaimonian relational values.
17 Here and in Lie (2016), chapter 4.5.3, I use 'adherence,' for what is usually called 'intrinsicness.'
18 Compare Holmes Rolston's notion that 'nature is a value producing system' (Rolston, 1986).
19 In *this* sense, we may argue that there is not necessarily a conflict between the classical ethical traditions (deontological, virtue and utilitarian ethics). One may see these traditions as 'ways to protect relations.' Utilitarianism (broadly speaking) might be an exception in that any relation can be broken for the sake of the utility of the parts. That is, relations are seen as contingent; feelings (pain and pleasure) or preferences located in the 'proper part' are not.
20 A useful analytical term for this 'downward influence' is 'demergence.' See Anjum and Mumford (2017).
21 See also Mathews, 1991, 129 ff. for a discussion on this issue.
22 Another interesting line of thought to consider is what is broadly referred to as the (weak) anthropic cosmological principle (e.g., Barrow & Tipler, 1986, *The Anthropic Cosmological Principle*, Oxford University Press). The number of fine-tuned facts—such as constants and "laws"—that we rely on is so extensive and precise that we require a theory of multiverses to maintain the notion of randomness. Philosopher Phillip Goff has recently argued that we need to move beyond agnosticism on this issue (Goff, 2023). However, we would still need an additional argument for why this noncontingent order is considered good.
23 It is not always easy to detect that most relational theories are "empty." Take, for instance, Helen Barad's theory of entanglement and 'agential cuts.' We are entangled, but are these entanglements more than merely contingent?
24 From a poststructuralist perspective, the suggested content here—possible-N relations—will generally be viewed (or rather deconstructed) as a naturalization of the structure/order, which holds power as the only 'truth-maker.' I have used the term 'naturalized' several times in this book, and this footnote is intended to clarify its meaning in *this* context. Here, 'naturalized order' refers to justifying a social order (often hierarchical) by claiming that the subordinated relation, such as the subordination of women, is natural.
25 Other philosophers have also made the same distinction. See, e.g., Prichard (2002) and Ross (2002).
26 The Danish word 'fordring' (in *Den etiske fordring*) is translated to 'The Ethical Demand' which is not very accurate. The word call is better, but not perfect.
27 In the Greek context, a lingering question remains about whether the urge is truly voluntary; according to the Aristotelian account, one would fail as a human being if unable to respond correctly to this urge for perfection.

28 To be sure, I use the term balance here not to suggest that any metaphysical principle of balance is involved. It is merely employed to express the phenomenological experience of a demand. When something is demanded of you and you cannot respond, you feel 'out of balance,' leading to an 'urge for balance.'

29 The fact that the description is arrived at by transcendental reasoning does not "excuse" the fact that such claims are descriptive and not normative.

30 From here and onward, I will distinguish between relations and relationships in a particular way. Relation(s) are general. Relationships are the particular instantiation of the relation. I have a relation with oxygen (in the air) and a relationship with the oxygen in the air that surrounds me.

31 This is what Kierkegaard would refer to as the rules of the 'Spidsborger,' while Heidegger calls them the rules of 'Das Man.' They are what they are, yet they could certainly be different. They may, of course, be wrongfully naturalized (e.g., women need to wear a scarf), thus attracting attention from freedom fighters.

32 We may recall that this is precisely what makes the right to ownership a fundamental right, according to John Locke.

33 As most scholars in this field know, it is rare to find someone who believes in limited economic growth or the idea that we should decide at a certain point to stop pursuing economic growth, even if this might seem like a good idea. A theory of economic growth typically does not include a section on how to halt growth and transition to a state without it. While no one argues that economic growth is synonymous with eternal growth, we may safely assume this is the prevailing belief.

34 I rely on the distinction between science, social science and the humanities, which is particularly relevant in the Anglo-American context.

35 Alternatively, we have seen that one might adopt a Humean perspective, asserting that findings in the laboratory are contingent. Neo-Humeans may employ statistics or discuss possible worlds to obscure Hume's radical, essentially nominalist view of nature. Regardless, the underlying idea is that the world consists of distinct parts that can be arranged in various ways (Molnar, 2003, 181). '*Anything may produce anything* else.' In Lie, 2016 (chapter 2), I argue that there is an *unholy* alliance between essentialism and Humeanism (see especially chapter 2.7).

36 Recall that reciprocity in this context means you must investigate other relational partners to understand what the partner you are examining is likely to do. This requires a back-and-forth analysis between the partnership and its manifestations, allowing *for inferences*. A partner's actions cannot be observed directly, as all manifestations are expressions of multiple dispositions. This underscores why science cannot rely solely on empirical studies and why we need theories; the ultimate reason being that nature is relational at its core.

37 "Overshoot means to go beyond limits inadvertently, without intending to do so" (Meadows et al., 1991, 1).

38 This claim requires further qualification. First, it is based on a liberal idea of freedom, which can vary across different legislative regimes. As previously discussed, the liberal notion of freedom asserts that individuals can do whatever they wish as long as their actions do not impede the freedom of others. The emphasis is on "doing whatever I want," meaning it is the responsibility of *the other* to demonstrate that they are being hampered. The burden of proof does not lie with the "instigator." Therefore, it is not Bayer's responsibility to prove that glyphosate has potential adverse effects, and they are free to challenge and question these concerns. A scientific perspective that initially interprets causal connections in this manner will inevitably struggle with *proving* those connections, which is fundamentally required. Is it glyphosate, or is it something else?

39 Other explanations of economic growth emphasize different aspects of "the equation," such as interest rate systems, the dynamics of capitalist competition and the autonomy of the financial system.

40 I hesitate when I use terms like "connecting services" or "connecting bureaucracies" because they imply an ontological distinction between "the technology itself" and its relationships, which contradicts a relational realist view. However, I make an exception here, trusting that the reader will keep this basic assumption in mind as I make this distinction for pedagogical reasons. The term "connective systems" arises from the need for further "coordination" of the possible-I systems that are established. The systems created generate problems (as Tainter & Patzek, 2012, 81) that must be addressed through connective efforts. We may refer to this as allo-connective systems, meaning other-connective systems.

41 See, e.g., Acemoglu & Robinson, 2012

42 This is how Joseph Tainter and Patzek articulate it:

> Confronted with problems, we often respond by developing more complex technologies (e.g., hybrid cars), establishing new institutions (e.g., the Department of Homeland Security), adding specialists or bureaucratic levels to an institution, increasing organization or regulation, or gathering and processing more information. Although we usually prefer not to bear the cost of complexity, our problem-solving efforts are powerful *complexity* generators. All that is needed for the growth of complexity is a problem that requires it. Problems continually arise, thus there is persistent pressure for complexity to increase.
>
> (Tainter & Patzek, 2012, 81)

References

Acemoglu, D., & Robinson, J. A. (2012). *Why nations fail: The origins of power, prosperity, and poverty*. Crown Business.

Anjum, R. L., & Mumford, S. (2017). Emergence and demergence. In M. Paoletti, and F. Orilia (Ed.), *Philosophical and scientific perspectives on downward causation* (pp. 92–109). https://doi.org/10.4324/9781315638577

Baard, P. (2024). Relational values is neither a necessary nor justified ethical concept. *Environmental Ethics*. https://doi.org/10.1080/21550085.2024.2334631

Barrow, J. D., & Tipler, F. J. (1986). *The anthropic cosmological principle*. Oxford University Press.

Brennan, A., & Lo, Y.-S. (2021). Environmental ethics. In E. N. Zalta (Ed.), *The Stanford Encyclopedia of philosophy* (Summer 2021 ed.). https://plato.stanford.edu/archives/sum2021/entries/ethics-environmental/

Brown, T. C. (1984). The concept of value in resource allocation. *Land Economics*, *60*(3), 231–246.

Callicott, J. B. (1985). Intrinsic value, quantum theory, and environmental ethics. *Environmental Ethics*, *7*(3), 257–275. https://doi.org/10.5840/enviroethics19857334

Carter, N. (2007). *The politics of the environment: Ideas, activism, policy* (2nd ed.). Cambridge University Press.

Chan, K. M. A., Gould, R., & Pascual, U. (2018). Editorial overview: Relational values: What are they, and what's the fuss about? *Current Opinion in Environmental Sustainability*, *35*, A1–A7. https://doi.org/10.1016/j.cosust.2018.11.003

Coeckelbergh, M. (2012). *Growing moral relations: Critique of moral status ascription*. Palgrave Macmillan.

Corning, P. A., et al. (2023). *Evolution "on purpose": Teleonomy in living systems*. The MIT Press.

Goff, P. (2023). *Why: The purpose of the universe*. Oxford University Press.

Hagens, N. J. (2020). Economics for the future – Beyond the superorganism. *Ecological Economics*, *169*. https://doi.org/10.1016/j.ecolecon.2019.106567

Hannis, M. (2017). *Freedom and environment: Autonomy, human flourishing and the political philosophy of sustainability*. Routledge.

Himes, A., & Muraca, B. (2018). Relational values: The key to pluralistic valuation of ecosystem services. *Current Opinion in Environmental Sustainability*, *35*, 1–7. https://doi.org/10.1016/j.cosust.2018.09.005
Kant, I. (1990). *Critique of pure reason*. Trans. N. K. Smith. Macmillan Education LTD. (Original work published in 1781).
Kant, I. (2015). *Critique of practical reason*. Trans. M. Gregor. Ed. A. Reath. Cambridge University Press. (Original work published in 1788).
Krieger, M. H. (1973). What's wrong with plastic trees? *Science*, *179*(4072), 446–455. https://doi.org/10.1126/science.179.4072.446
Lee, K. (2003). The source and locus of intrinsic value: A reexamination. In A. Light & H. Rolston, III (Eds.), *Environmental ethics: An introduction*. Blackwell Publishing.
Lie, S. A., & Wickson, F. (2011). The relational ontology of deep ecology: A dispositional alternative to intrinsic value. In A. Aaro & J. Servan (Eds.), *Environment, embodiment & history*. Hermes Texts.
Lie S.A.N. (2016). *Philosophy of nature: Rethinking naturalness*. Routledge.
Løgstrup, K. E. (2020). *The ethical demand*. Trans. B. Rabjerg & R. Stern. Oxford University Press. https://doi.org/10.1093/oso/9780198855989.001.0001
Maier, D. S., & Feest, A. (2016). The IPBES conceptual framework: An unhelpful start. *Journal of Agricultural and Environmental Ethics*, *29*, 327–347. https://doi.org/10.1007/s10806-015-9584-5
Mathews, F. (1991). *The ecological self*. Routledge.
Meadows, D. H., Meadows, D. L., & Randers, J. (1991). *Beyond the limits: Confronting global collapse, envisioning a sustainable future*. Chelsea Green Publishing.
Molnar, G. (2003). *Powers: A study in metaphysics*. Oxford University Press.
Muraca, B. (2016). Relational values: A Whiteheadian alternative for environmental philosophy and global environmental justice. *Balkan Journal of Philosophy*, *8*(1), 19–30.
Næss, A. (1973). The shallow and the deep, long-range ecology movement: A summary. *Inquiry: An Interdisciplinary Journal of Philosophy*, *16*(1–4), 95–100.
Næss, A. (1989). *Ecology, community and lifestyle: Outline of an ecosophy*. Cambridge University Press.
Næss, A. (2002). *Life's philosophy: Reason and feeling in a deeper world*. University of Georgia Press.
Noddings, N. (1984). *Caring: A feminine approach to ethics and moral education*. University of California Press.
Norton, B. G., & Sanberg, J. (2021). Relational values: A unifying idea in environmental ethics and evaluation? *Environmental Values*, *30*(6), 695–714. https://doi.org/10.3197/096327120X16033868459458
O'Neill, J. (1992). Varieties of intrinsic value. *The Monist*, *75*(2), 119–137.
O'Neill, J. (2010). The overshadowing of needs. In F. Rauschmayer, I. Omann, & J. Frühmann (Eds.), *Sustainable development: Capabilities, needs, and well-being* (pp. 25–42). Routledge.
Palmer, C. (2003). An overview of environmental ethics. In A. Light & H. Rolston, III (Eds.), *Environmental ethics: An anthology*. Blackwell Publishing.
Pescual, U, et al. (2018). Sustainability challanges: Relational values. In *Current opinion in environmental sustainability* (pp. 1–132).
Prichard, H. A. (2002). Duty and interest. In T. MacAdam (Ed.), *Moral writings*. Oxford University Press (Originally published in 1928).
Rauschmayer, F., et al. (2010). Needs, capabilities, and quality of life: Refocusing sustainable development. In F. Rauschmayer, I. Omann, & J. Frühmann (Eds.), *Sustainable development: Capabilities, needs, and well-being* (pp. 1–25). Routledge. https://doi.org/10.4324/9780203839744

Rolston, H. III. (1986). *Environmental ethics: Duties to and values in the natural world*. Temple University Press.
Rolston, H. I. I. I. (2003). Value in nature and the nature of value. In A. Light & H. Rolston, III (Eds.), *Environmental ethics: An introduction*. Blackwell Publishing.
Ross, W. D. (2002). *The right and the good*. Ed. P. Stratton-Lake. Oxford University Press. (Originally published in 1930).
Scavenius, T. (2018). Climate change and moral excuse: The difficulty of assigning responsibility to individuals. *Journal of Agricultural and Environmental Ethics*, *31*(1), 1–15. https://doi.org/10.1007/s10806-018-9705-z
Stephens, P. (2001). The green only blooms among the Millian flowers: A reply to Marcel Wissenburg. *Environmental Politics*, *10*(3), 43–47. https://doi.org/10.1080/714000563
Tainter, J. A., & Patzek, T. W. (2012). *Drilling down: The Gulf oil debacle and our energy dilemma*. Springer.
Taylor, C. (1992). The politics of recognition. In A. Gutmann (Ed.), *Multiculturalism and "The politics of recognition"* (pp. 25–73). Princeton University Press.
Torres, O. (2022). Expansion of humanity into space: Utopia and dystopia. In C. S. Cockell (Ed.), *The institutions of extraterrestrial liberty*. Oxford University Press.
Toulmin, S. (1990). *Cosmopolis: The hidden agenda of modernity*. The University of Chicago Press.
Vogel, S. (2015). *Thinking like a mall*. MIT press.
Zimmer, C. (2013). Species revival: Should we bring back extinct animals? *National Geographic Magazine*. https://www.nationalgeographic.com/magazine/article/species-revival-bringing-back-extinct-animals
Zubrin, R. (2022). The case for space is liberty. In C. S. Cockell (Ed.), *The institutions of extraterrestrial liberty*. Oxford University Press.

Index